Future Oil Demands of China, India, and Japan

Future Oil Demands of China, India, and Japan

Policy Scenarios and Implications

George G. Eberling

LEXINGTON BOOKS
Lanham • Boulder • New York • London

Published by Lexington Books
An imprint of The Rowman & Littlefield Publishing Group, Inc.
4501 Forbes Boulevard, Suite 200, Lanham, Maryland 20706
www.rowman.com

16 Carlisle Street, London W1D 3BT, United Kingdom

British Library Cataloguing in Publication Information Available

Library of Congress Cataloging-in-Publication Data
Eberling, George.
Future oil demands of China, India, and Japan : policy scenarios and implications / George G. Eberling.
pages cm
Includes bibliographical references and index.
ISBN 978-0-7391-9181-1 (cloth : alk. paper) — ISBN 978-0-7391-9182-8 (electronic : alk. paper)
1. Petroleum industry and trade—China. 2. Petroleum industry and trade—India. 3. Petroleum industry and trade—Japan. 4. Energy consumption—China. 5. Energy consumption—India. 6. Energy consumption—Japan. 7. Petroleum reserves—China. 8. Energy policy—China. 9. China—Foreign economic relations—India. 10. India—Foreign economic relations—China. 11. China—Foreign economic relations—Japan. 12. Japan—Foreign economic relations—China. I. Title.
HD9576.C52E236 2014
333.8'23212095—dc23 2014015734

™ The paper used in this publication meets the minimum requirements of American National Standard for Information Sciences—Permanence of Paper for Printed Library Materials, ANSI/NISO Z39.48-1992.

Printed in the United States of America

Contents

List of Tables

List of Figures

List of Abbreviations

ARF	ASEAN Regional Forum
ASEAN	Association of South East Asian Nations
BAP	Beijing Action Plan
BBL	Oil Barrels Per Year
BRIC	Brazil, Russia, India, China
CBM	Confidence Build Measures
CBO	Congressional Budget Office
CNG	Compressed Natural Gas
CNOOC	China National Offshore Oil Corporation
CNPC	China National Petroleum Corporation
CRS	Congressional Research Service
CSIS	Center for Strategic and International Studies
CTF	Combined Task Force
CUPET	Cubapetroleo
CVBG	Carrier Battle Group
CVN	Nuclear Aircraft Carrier
DSI	Diaoyu/Senkaku Islands
EEZ	Economic Exclusion Zone
EIA	Energy Information Administration
ESPO	East Siberian Pacific Ocean
EU	European Union
EXIM	Export-Import
FDI	Foreign Direct Investment
FOCAC	Forum on China-Africa Cooperation
HEV	Hybrid Electric Vehicle
IEA	International Energy Agency
IMF	International Monetary Fund

IOC	Independent Oil Company
JAPEX	Japanese Petroleum Exploration Co. Ltd.
KPI	Kuwait Petroleum International
MEND	Movement for the Emancipation of Niger Delta
MES	Market Economy Status
MOOTW	Military Operations Other Than War
MPLA	Popular Movement for the Liberation of Angola
NAM	Non Aligned Movement
NATO	North Atlantic Treaty Organization
NOC	National Oil Company
NORINCO	China North Industries Corporation
ODA	Official Development Assistance
OECD	Organization of Economic Co-operation and Development
OIL	Oil India Limited
ONGC	Oil and National Gas Corporation
OPEC	Organization for Petroleum Exporting Countries
OVL	ONGC Videsh Limited
PKO	Peacekeeping Operations
PLA	People's Liberation Army
PLAN	People's Liberation Army Navy
PRINCE	Probe, Interact, Calculate, and Execute
SAGIA	Saudi Arabian General Investment Authority
SCO	Shanghai Cooperation Organization
SEZ	Special Economic Zone
SINOPEC	China Petroleum and Chemical Corporation
SIPRI	Stockholm International Peace Research Institute
SLOC	Sea Lines of Communication
SSBN	Nuclear Ballistic Missile Submarine
SSGN	Nuclear Guided Missile Submarine
SSN	Nuclear Attack Submarine
SPR	Strategic Petroleum Reserve
TICAD	Tokyo International Conference on African Development
UAE	United Arab Emirates
UN	United Nations
UNITA	National Union for the Total Independence of Angola
WMD	Weapons of Mass Destruction

I

THE DEPENDENCY SCENARIO

1

Introduction

In *Chinese Energy Futures and Their Implications for the United States*, I examined how China's oil energy needs will likely shape future Sino-American relations under the conditions of dependency or non-dependency. Using scenario analysis and the PRINCE (Probe, Interact, Calculate, and Execute) forecasting system to estimate the likelihoods of three oil energy futures or scenarios (competitive dependency, competitive surplus, and cooperative surplus), I discussed and evaluated the strategic implications of these energy futures for the United States and determined the most likely oil energy future.

In this book, I build upon this analysis to examine how China's oil energy needs will likely shape future Sino-Indian relations and Sino-Japanese relations. Obviously, as China's thirst for oil grows, so will its political and economic status and influence. Militarily, it is very likely that Chinese naval presence and activity may increase not only in the East China Sea and South China Sea but also in the Indian Ocean and the Persian Gulf; geographic areas which have strategic political, economic and military significance to India and Japan. Beyond the strategic maritime commons, political and economic competition for oil energy resources may evolve on land in the Near East, Sub-Saharan Africa, and the Western Hemisphere; distinct geographic areas which are rich in natural energy resources.

Given that crude or refined petroleum is a finite strategic commodity, that energy demand from non-OECD (Organization for Economic Co-operation and Development) countries like China and India is likely to overtake that of OECD countries like Japan over the next twenty years,[1] that China may be the world's largest energy consumer of oil in the next twenty years, that non-OPEC (Organization of Petroleum Exporting

Countries) oil production is predicted to plateau by 2030,[2] thereby increasing the Near East's strategic importance, that China, India, and Japan largely depend on foreign petroleum resources to sustain their national economies, that India and Japan are the two largest consumers of oil after China, and that India and Japan are political and military rivals of China in Asia, it is important to know how China's insatiable thirst for petroleum energy resources will affect its relations in terms of either regional and/or global influence with its two largest energy competitors after the United States: India and Japan.

Oil also has a geopolitical dimension. According to Philip Andrews-Speed and Roland Dannreuther:

> Oil is different from other commodities in the extent to which its production, supply and distribution are marked by geopolitical tensions and sensitivities. To a degree surpassing other commodities, oil excites considerable strategic nervousness and suspicion. And as China is driven to secure ever-increasing quantities of oil from different parts of the world, so anxieties and suspicions of other countries about China's intentions and ambition will grow.[3]

In the international context, Philip Andrews-Speed and Roland Dannreuther opined oil's speciality derives from a number of features:

> Its physical properties make it a liquid fuel with a high quantity of energy per unit of volume and of weight, and ideally suited to past and current technologies for transport; conventional geological resources of crude are becoming increasingly concentrated in a small number of countries, principally in the Middle East, North Africa and the former Soviet Union, and its fungible nature and relative ease with which it can be transported around the world have resulted in a market for oil that is essentially global in nature, through regional hubs.[4]

HISTORICAL OVERVIEW OF CHINA'S OIL ENERGY DEPENDENCY

China was oil energy resource self-sufficient until 1993 when its domestic oil production began to lag behind its growing oil consumption and the country became a net importer of oil.[5] Figure 1.1 illustrates China's daily oil consumption versus production since 1980 including 1993 when it became dependent on foreign petroleum.

Since then, China has been securing petroleum energy resources globally through the acquisition of production sources and facilities in addition to oil products, by actively seeking concessions for oil exploration and production and by establishing energy relationships.[6]

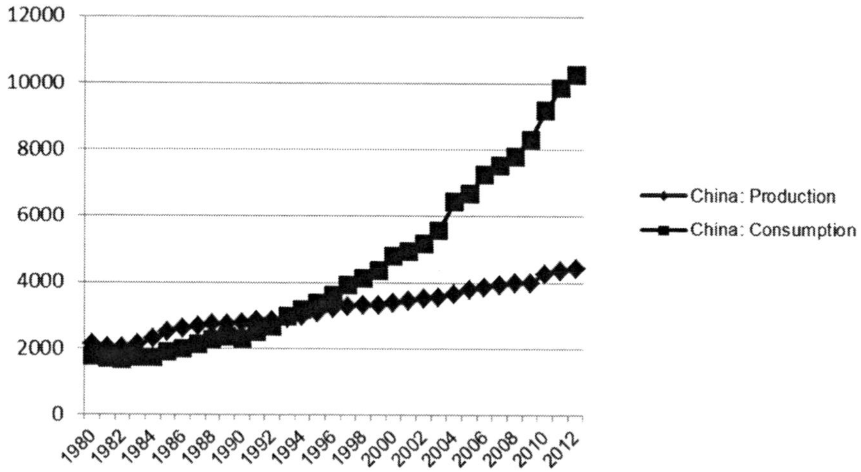

Figure 1.1 China's Daily Oil Consumption vs. Production (1980–2012)

To date, China imports petroleum from fifty-six countries and has oil exploration and production operations in forty-three countries. In comparison, India and Japan import oil from twenty-eight and twenty countries respectively and have oil exploration and production operations in seventeen and twenty-five countries respectively. See Table 1.1. Those countries in bold print export crude oil to China, India, and Japan. India and Japan have been exploring and producing oil as well as importing oil long before China became a net oil importer in 1993. Yet, China has been more aggressive in both respects in the last two decades. In the last twenty years, China has been tapping into India and Japan's principal oil supplies from Angola, Nigeria, South Sudan, Venezuela, Iran, Kuwait, Saudi Arabia, United Arab Emirates (UAE), and Russia.

Of particular significance is China's failed attempt to acquire the U.S. energy company Unocal in 2005 for $18.5 billion in cash. While unsuccessful and not directly related to energy resource completion between China and India or China and Japan, the Unocal incident clearly demonstrates China's aggressiveness in securing energy resources globally. Further, the incident clearly indicates that energy resource competition does exist between the world's great powers. If China is willing to compete with the United States for petroleum resources, it will surely compete with India and Japan as well, especially since more than 50 percent of China, India, and Japan's total annual petroleum consumption is from imported oil. Japan is wholly resource dependent on foreign oil because it has virtually no natural resources produced at home. Thus, Japan must import almost 100 percent of its oil to sustain its economy and military.

Table 1.1 China, India, and Japan's Oil Suppliers

Sub-Saharan Africa	Western Hemisphere	Near East	Asia Pacific	Europe, Eurasia and Russia
Angola	Argentina	**Algeria**	**Australia**	**Azerbaijan**
Cameroon	Brazil	**Iran**	**Brunei**	Norway
Chad	Canada	**Iraq**	Burma	**Russia**
Congo, Republic of the	Colombia	**Kuwait**	Indonesia	United Kingdom
Equatorial Guinea	Cuba	Libya	**Malaysia**	
Gabon	**Ecuador**	**Oman**	Mongolia	
Mauritania	Peru	**Qatar**	Philippines	
Nigeria	**Venezuela**	**Saudi Arabia**	Thailand	
South Africa		**United Arab Emirates**	Vietnam	
South Sudan		**Yemen**		

Source: FACTS Global Energy.

For comparison purposes, figures 1.2 and 1.3 illustrates India and Japan's daily oil consumption versus production since 1980 in 1,000 barrels per day. The difference between consumption and production in these figures indicates the amount of average crude oil that must be imported daily to sustain India and Japan's economy and military. Naturally, crude oil must be refined into petroleum.

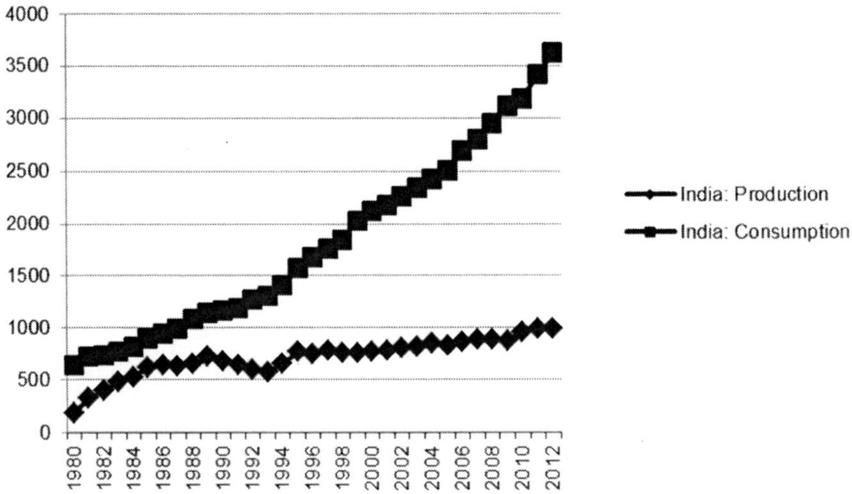

Figure 1.2 India's Daily Oil Consumption vs. Production (1980–2012)

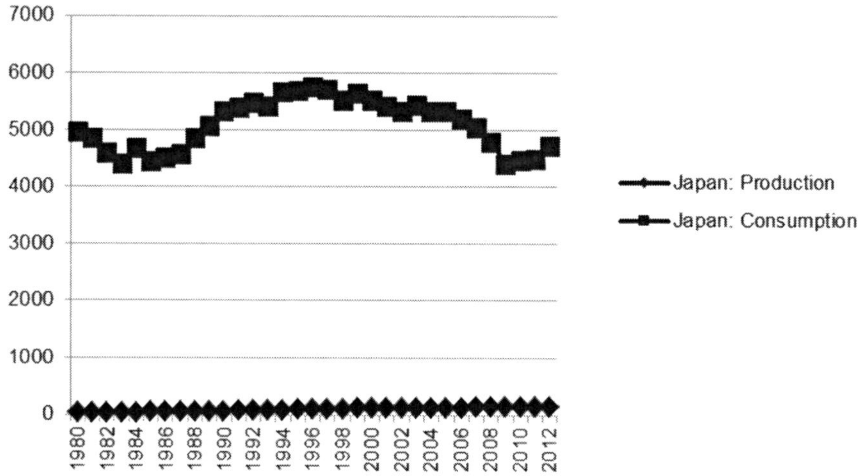

Figure 1.3 Japan's Daily Oil Consumption vs. Production (1980–2012)

Despite China's rising use of oil, great potential exists in future energy savings in China's transport sector through further raising fuel efficiency standards in vehicles, replacing old vehicles, changing the structure of the road fleet, by encouraging the use of diesel and hybrid cars, and by investing in urban transport systems.[7] In this context, China has taken a number of steps to constrain the rise of oil use in this sector since 2006:

> It has raised the level of purchase tax on large cars, raised the fuel economy standards for vehicle manufacturers, increased state funding for research and development in alternative-fuelled vehicles, and ordered that government departments use vehicles with higher fuel efficiency standards and that departmental vehicles not be available for private use by employees.[8]

However, the argument for an energy efficient and environmentally friendly transport policy seems to have been subsumed beneath an industrial policy that has championed automobile manufacturing, an urban design policy that has favored multi-lane highways as the main transport network, and a social policy that, intentionally or otherwise, has highlighted car ownership as a legitimate expectation of the growing middle classes.[9]

THREE OIL ENERGY FUTURES OR SCENARIOS

China's desire to pursue global oil competitively in the wake of its crude oil and petroleum dependency raises some important questions for India and Japan. Of great importance, what will be the strategic implications of China's oil energy needs for India and Japan under current conditions, that is, a scenario of competitive dependency? The competitive nature of Sino-Indian and Sino-Japanese energy needs could potentially lead in two directions: competition for scarce oil resources or cooperation for those scarce resources. Further, what are the strategic implications of China's oil energy needs for India and Japan under other conditions such as a competitive surplus (non-dependency) scenario or a cooperative surplus (non-dependency) scenario? In other words, how will each energy future or scenario under the conditions of dependency and non-dependency affect India and Japan globally on a political, economic and military scale?

Other questions to be asked include:

- What events will be associated for each oil energy future or scenario?
- What will be the most likely Chinese oil energy future or scenario?
- What will be the most likely policy responses of India and Japan?

It is essential that these questions and more be fully addressed and answered because the absence of any definitive conclusions on China's oil energy future under the conditions of dependency and non-dependency presents a significant gap in the existing body of knowledge with respect to global energy security trends. This book will hold significant implications for policymakers and government officials alike and will help them to make more informed foreign policy decisions, especially those that influence or affect global energy security. To that end, three scenarios are analyzed to determine how they determine the most likely public policy course between China, India, and Japan as they vie for current and future oil sources:

- A *competitive dependency scenario* is one in which China imports foreign crude oil and petroleum to sustain economic growth, competition exists for global oil supplies between China and India and China and Japan (China, India, and Japan import crude and refined petroleum from the same producers) and competition exists between China and Brunei, Malaysia, the Philippines, Taiwan, and Vietnam in the South China Sea region;
- A *competitive surplus scenario* is one in which China relies heavily on fuel-saving policies and technologies (hybrid, electric, compressed natural gas) to decrease consumption and increase production, and competition exists between China and Brunei, Malaysia, the Philippines, Taiwan, and Vietnam with respect to oil exploration, extraction, refinement, and transportation in the South China Sea region.

China explores for oil in contested territorial waters, and China relies predominantly on domestically produced oil and;

- A *cooperative surplus scenario* is one in which China relies heavily on fuel-saving policies and technologies (hybrid, electric, compressed natural gas) to decrease consumption and increase production, and cooperation exists between China and Brunei, Malaysia, the Philippines, Taiwan, and Vietnam with respect to crude oil exploration, extraction, refinement, and transportation in the South China Sea region. China explores for oil in resolved territorial waters, and China relies predominantly on domestically produced oil.

In the competitive dependency scenario, China uses a "Go Out" policy or "Going Global" strategy to secure energy resources abroad. This policy seeks to employ China's foreign reserves by acquiring assets overseas and equipping its domestic firms (such as China's national oil companies) and their management with international experience so that they can take the competition to the home markets of foreign nations and compete better in mainland China's own domestic market. China National Offshore Oil Corporation's (CNOOC) $18.5 billion proposed purchase of Unocal is illustrative of the "Go Out" policy in which the Chinese government (a state actor) provided the financial leverage for CNOOC to purchase Unocal.

In the non-dependency scenarios (competitive surplus and cooperative surplus) China does not use a "Go Global" strategy because China does not import foreign oil. It is not necessary to use a "Go Out" policy because China uses fuel-saving technologies and domestic oil resources.

The scope of this book is limited to three possible Chinese oil energy futures and their strategic implications for India and Japan using scenario analysis because this perspective paints the broadest picture capturing the full range of potential alternative futures even though there will be permutations and combinations of the three futures.[10] While other energy futures offer considerable merit and attention to address China's dependency of foreign crude oil and refined petroleum such as scenarios based on the use of biofuel technologies (ethanol and/or manure processing) or other "unconventional" fossil fuel resources such as heavy oil, tar sands, oil shale or gas resources—which are not counted at proven oil reserves[11]—such scenarios would involve exhaustive qualitative and quantitative research and data that would extend well beyond the scope of the investigation. My research leads me to believe that these are very low probability events and thus futile to explore. Further, such futures would deviate considerably from what this research seeks to accomplish in the realm of possible oil energy futures in light of estimated levels of global oil production, oil consumption, and proven oil reserves. In this respect, Iain Conn of British Petroleum estimates that the world has approximately forty-one years of conventional crude oil reserves remaining.[12] This book seeks only to examine and determine those energy futures that have the highest probability of occurrence and the significance for future Sino-Indian relations and Sino-Japanese relations (see table 1.2 below).

I considered using a cooperative dependency oil energy future for this book in which China was dependent on foreign oil to sustain economic growth but did not import oil from the same oil producers as India or Japan. However, China would not be able to sustain economic growth in this future because it would not be able to meet its annual oil energy consumption demands. In this respect, China imports almost 50 percent (at least 500 million barrels per year) of its oil from the same countries as India and Japan.

Table 1.2 Three Chinese Oil Energy Futures or Scenarios

Competitive Dependency Scenario
Oil production lags behind consumption.
Competition exists for global oil supplies between China, India, and Japan (China, India, and Japan import crude oil and petroleum from the same producers).
China imports foreign crude oil and petroleum to sustain economic growth. Competition exists between China and Brunei, Malaysia, the Philippines, Taiwan, and Vietnam with respect to oil exploration, extraction, refinement, and transportation in the South China Sea.
Competitive Surplus Scenario (Non-Dependency)
Oil production matches or exceeds consumption.
Competition exists between China and Brunei, Malaysia, the Philippines, Taiwan, and Vietnam with respect to oil exploration, extraction, refinement, and transportation in the South China Sea.
China explores for oil in contested territorial waters; China relies predominantly on domestically produced oil.
China relies heavily on fuel-saving policies and technologies (hybrid, electric, compressed natural gas) to decrease consumption and increase production.
Cooperative Surplus Scenario (Non-Dependency)
Oil production matches or exceeds consumption.
Cooperation exists between China and Brunei, Malaysia, the Philippines, Taiwan, and Vietnam with respect to crude oil exploration, extraction, refinement, and transportation in the South China Sea.
China explores for oil in resolved territorial waters; China relies predominantly on domestically produced oil.
China relies heavily on fuel-saving policies and technologies (hybrid, electric, and compressed natural gas) to decrease consumption and increase production.

Because China's daily oil consumption exceeded daily oil production since 1993 and almost doubled in 2008, it would have to implement draconian public policy measures to sustain its economic growth and make up for a 50 percent loss in oil imports under the cooperative dependency scenario. Such policies would include but not be limited to decreasing oil consumption and increasing oil production such as petroleum rationing or tapping into its strategic petroleum reserve (SPR). China would have to consider fuel-saving automotive technologies such as hybrid electric vehicles (HEVs), compressed natural gas (CNG) vehicles or even fuel cell vehicles. Because oil is critical to China's economic growth, it would be unrealistic to expect China to not import oil from the same oil producing countries as India or Japan.

Table 1.3 provides a graphic representation of the dependency oil energy futures (competitive dependency and cooperative dependency) and the non-dependency oil futures (competitive surplus and cooperative surplus). Note that the cooperative dependency oil energy future was not considered.

Table 1.3 Chinese Oil Energy Futures or Scenarios

Oil Energy Future/Scenario #1: *Competitive Dependency* (Oil Dependency)	**Oil Energy Future/Scenario #2:** *Competitive Surplus* (Oil Non-Dependency)
Oil Energy Future/Scenario #4: *Cooperative Dependency* (Oil Dependency) *This future/scenario was not considered.*	**Oil Energy Future/Scenario #3:** *Cooperative Surplus* (Oil Non-Dependency)

SUPPORTING ARGUMENTS

Several arguments help to inform our understanding of future Sino-Indian and Sino-Japanese relations with respect to oil resources. The first argument relates to the anxieties and growing danger of conflict associated with acquisition of energy resources; the second argument relates to "resource nationalism" which might be defined as the management of energy flows in accordance with vital state interests. It is noted that these arguments are based on literature that is primarily Sino-American centric because of the limited availability of literature on Sino-Indian energy relations except *China, India and the United States: Competition for Energy Resources* or Sino-Japanese energy relations. Nevertheless, they are helpful.

In *Blood and Oil: The Dangers and Consequences and America's Growing Dependency on Imported Petroleum, Resource Wars: The New Global Landscape of Global Conflict,* and *Rising Powers, Shrinking Planet: The New Geopolitics of Energy,* Michael T. Klare writes extensively about the problems associated with the acquisition of energy resources by the world's most consuming nations like China and the United States. Using case studies as evidence, he argues that petroleum production is in state of decline because the global energy industry will not be able to increase the output of crude oil to satisfy the ever-rising levels of consumption by large industrial nations and that global production capacity arises from a steady decline in the rate of new oil-field discovery. Klare contends that competition for access to oil among the world's major consuming nations (like China and the United States) is expected to become increasingly fierce in light of the fact that the global energy industry will not be able to ramp up production to satisfy the rising demand for oil. In addition, Klare argues that a shift has occurred in the center of gravity of world oil production from the developed world to the developing world, which has profound implications for energy-seeking nations because it entails a heightened risk to the uninterrupted flow of energy supplies.[13] Heightened risk of instability and violence in the oil-producing areas (due to the specific targeting of oil facilities by anti-Western extremists in recent times), will increase the risk of disruptions in the flow of oil from the supplying to consuming nations. Alarmed by this prospect, Klare anticipates that the United States will increasingly rely on military instruments to safeguard the flow of oil, which is known as the "militarization of energy management."[14]

On a similar vein, in the *Alliance Curse* and *Capital and Collusion,* Hilton Root employs case studies to write about dangers associated with asymmetric alliances between first and third world countries. He argues that throughout history, "an alliance between a greater power and a weaker strategically significant dictator traps the dominant partner into unnecessary conflicts and moral compromises."[15] Although Root does not refer specifically to oil as the source of the weaker nation's strength in the unlikely relationship or even China's asymmetric relations with the developing world in the post Cold War era, he demonstrates how these types of relationships undermine political reform, economic development, and the rule of law in the weaker nations of developing Asia and Latin America. Essentially, political and social instability results and economic development is stymied.

The case studies presented by Klare and Root are quite illustrative regarding the anxieties and dangers of first and third world relationships or "alliances" with large industrial nations. Klare goes a step further to discuss the specific anxieties and dangers associated with the acquisition of energy resources by world's most consuming nations like China and the United States. While Klare's work is intuitive in nature and qualitative in approach, qualitative analysis would enhance his work further.

In this respect, table 1.4 shows a direct relationship between the world's top ten oil exporters and China, India, and Japan, the exporting countries' average daily oil net exports (in 1,000 barrels per day), China, India, and Japan's average daily oil imports from each exporter (in 1,000 barrels per day) and the level of freedom associated with each oil exporter.

Table 1.4 Freedom in the World Compared with Average Daily Oil Exports and Average Daily Oil Imports by China, India, and Japan in 2012

Country of Origin	Freedom House Score	Net Exports (bbl)	China's Oil Imports (bbl)	% of China's Total Daily Imports (bbl)	India's Oil Imports (bbl)	% of India's Total Daily Imports (bbl)	Japan's Oil Imports (bbl)	% of Japan's Total Daily Imports (bbl)
Saudi Arabia	Not Free	8,865	1075.1	20%	589	16%	1,138	31%
Russia	Not Free	7,201	485	9%	3	0.1%	167	4.6%
Iran	Not Free	1,880	439	8%	315	9%	189	5%
United Arab Emirates	Not Free	2,595	174.4	3%	301	8.4%	792.6	22%
Norway	Free	1,684	0	0%	3.1	0.1%	0	0%
Kuwait	Partly Free	2,414	209.3	4%	303	8.5%	268	7%
Nigeria	Partly Free	2,254	18.7	0.3%	265	7.4%	10.5	0.3%
Angola	Not Free	1,178	800.9	15%	155	4.3%	7.7	0.2%
Iraq	Not Free	2,235	312.8	6%	473	13.2%	75.7	2%
Venezuela	Partly Free	1,712	305	5.6%	348	9.7%	5.6	0.2%
Total (2012)		32,018	3,820.2	80.9%	2,755	76.7%	2,654.1	72.3%

Source: Freedom House, Facts Global Energy and U.S. Department of Energy.

Freedom House reports that of the top ten oil exporting nations which exported oil to China, India, and Japan in 2012, 60 percent or six countries were not free (Angola, Iran, Iraq, Russia, Saudi Arabia, and United Arab Emirates), 30 percent or three countries were partly free (Kuwait, Nigeria, and Venezuela) and 10 percent or one country was free (Norway). Moreover, in terms of energy resource competition between

China, India, and Japan, it appears that at least 72 percent of their oil was imported from the same exporters: Angola, Iran, Iraq, Kuwait, Nigeria, Russia, Saudi Arabia, the United Arab Emirates, and Venezuela. China and India imported approximately 77 percent of their oil from the same exporters: Angola, Iran, Iraq, Kuwait, Nigeria, Saudi Arabia, the United Arab Emirates, and Venezuela. Further, China and Japan imported approximately 72 percent of their oil from the same exporters: Angola, Iran, Iraq, Kuwait, Nigeria, Russia, Saudi Arabia, the United Arab Emirates, and Venezuela.

In light of the fact that China, India, and Japan import at least 72 percent of their crude oil from socially and politically unstable countries, which potentially could result in supply disruptions, national economic disaster, and intensive competition for energy resources—it is necessary to determine empirically how and to what degree oil will influence this process using a reliable forecasting methodology.

RESEARCH DESIGN

Scenario analysis provides the best way to establish quantitative causality between China's alternative futures, or scenarios (and associated events), and the possible strategic implications for India and Japan. Scenario analysis is a process of analyzing possible future events by considering alternative possible outcomes.

> Through the development and use of scenarios, we challenge our assumptions and be better prepared for the unexpected when it, inevitably occurs.[16]
>
> Scenario analysis focuses on the areas of greatest uncertainty for a country or an operation, systematically develops several plausible alternative future environments in which the operation might be implemented, and determines how they would affect its success . . . The analysis is designed to allow improved decision-making by allowing more complete consideration of outcomes and their implications. . . . In politics or geo-politics, scenario analysis involves modeling the possible alternative paths of a social or political environment and possibly diplomatic and war risks.[17]

In addition, the PRINCE forecasting method that was developed by William D. Coplin and Michael K. O'Leary is used in this book to determine whether China and India and Japan will peacefully share the world's oil reserves or whether competition for oil and gas reserves will lead to conflict.

According to William D. Coplin and Michael K. O'Leary:

> The PRINCE model "is useful in a wide variety of political situations, and is one of the few approaches to political forecasting that is both systematic and relevant to decision-makers. The model provides a means for the rigorous analysis of information, producing calculations that result in forecasts of the probability of specific outcomes. At the same time, it uses the invaluable and unique expertise of specialists—both to supply information about the model and to adjust the calculations produced by the model according to qualitative, subjective knowledge about a given country."[18]

In the PRINCE model, the actors (or stakeholders) are identified by their power, position and salience, which are used to calculate the probabilities for each scenario's associated outcome, determining which energy future the Chinese are most likely to adopt. Stakeholders are identified based on the existing body of literature, their domestic and international political relationships and interactions with other stakeholders, and the value of their available resources, which equates to their relative power, and their position (stance) on the issue of Chinese oil futures and their salience for the issue of oil competition given all other important issues that each stakeholder must deal with, will be estimated. These factors are assigned a numerical value based on the perspectives of survey participants who have knowledge of Chinese oil energy politics. Surveys were sent to over four hundred experts in the United States, Asia, and Western Europe. Survey participants were composed of individuals from academia, government, and the private sector who have specialized knowledge of any of the following: oil politics and/or economics, Chinese oil energy politics, Chinese politics and/or economics, Chinese maritime security issues, the comparative politics of Asia, Africa, Europe (including Russia), Latin America, North America, and the Near East as it relates to oil or national security issues.

Event identification and assignation to a particular Chinese oil energy scenario is determined by actual and potential world events and whether the event directly involves the supply of oil from its country or region of origin or indirectly involves the supply of oil through a third country or through a national security event.

In his seminal work *Expert Political Judgment: How good is it? How can we know?*, Philip Tetlock (2005) found that experts in foreign affairs make poor predictions about future events and were no better forecasters than informed non-experts in making predictions. Tetlock also observed that experts tend to overestimate the probability of likely events that will occur as well as overestimate the probability of unlikely events that will not occur, resulting in less accurate predictions of future events. As well, in "How Statistical Models Can Help Intelligence Analysts," Steven Rieber (2004) observed that political experts tend to misjudge not only which factors (causes and effects) are relevant but how they are relevant.

He observed that experts will often fail to predicate their prediction estimates on a large sample of relevant data, relying instead on small numbers of memorable or anecdotal illustrations. Reiber suggested that formal models can usefully supplement expert judgment because they test the validity of factors/variables and their assigning weights; base their predictions on all known relevant data such as statistical samples; minimize both random and systematic error; establish useful baselines; deepen experts' understanding of key entities, classes, concepts, and relationships; and reveal underlying causality in phenomena.

Formal models are also easy to use but require data. Experts can and do provide data. When expert data is used and properly calibrated, more accurate predictions of future outcomes will result. Otherwise, in the absence of using expert data and a formal model for forecasting futures such as the PRINCE model, experts will tend to overestimate the probability of likely events that will occur as well as overestimate the probability of unlikely events that will not occur, resulting in less accurate predictions of future events.

THE MECHANICS OF THE PRINCE MODEL

PRINCE is an acronym for four steps: "**Pr**obe, **In**teract, **C**alculate, and **E**xecute." Each step is integral to the PRINCE model of risk analysis forecasting. The following discussion is indebted to William Coplin, Pat McGowan, and Michael O'Leary's *Everyman's Prince: A Guide to Understanding Your Political Problems* (1972).

Step 1: Probe

In the probe step policy experts define the issue and identify the actors. An issue is a proposed decision or action which is likely to generate controversy. The policy experts surveyed should include a cross-range of theorists and practitioners who have first-hand knowledge of the issue at hand. An actor is any individual, group, or organization that ought to be considered in making the decision or in carrying it out after it has been made. This information is necessary to analyze each actor's influence on the issue.

Step 2: Interact

Once the actors and issues have been identified by policy experts the next step is to analyze each actor's *influence* on the issue. Influence is the combination of position, salience, and power. These three variables are necessary to calculate the actors' influence.

- *Position* is whether one favors or opposes a particular solution to an issue. Power is measured on a scale of -5 to +5 (-5, -4, -3, -2, -1, 0, 1, 2, 3, 4, and 5). The higher the number, the greater an actor favors (+) or opposes (-) a particular solution to an issue.
- *Salience* is the "strength" of the support or opposition. Salience is measured on a scale of 1 to 5 (1, 2, 3, 4, and 5). The higher the number, the greater the "strength" of the support or opposition of the decision-making actor.
- *Power* is the ability to make one's preference on an issue take place. Power is measured on a scale of 1 to 5 (1, 2, 3, 4, and 5). The higher the number, the greater the actor's ability to actualize his/her preference.

Policy experts estimated each actor's position, salience, and power relative to the other actors making the decision. Each actor's relative influence (the combination of position, salience, and power) is then recorded on a matrix (see step 3 below). Ideally, a sample size of at least thirty policy experts is recommended to participate in the PRINCE survey.

Step 3: Compute

Once information has been gathered on each actor's position, salience, and power, the next step is to build a series of tables summarizing the findings. This is a two-step process of determining each actor's weight and calculating the probability of the issue (decision) being adopted.

Position, Salience, and Power Matrix: The actors are listed in the first column and the policy issue across the top of the matrix and the numerical values entered for each factor. Actors with no position (either no position yet or a "firm" I don't care) on an issue are assigned a "0" for that issue. The position variable may be positive (+) or negative (-) while the other variables (power and salience) are always positive. The sum of the actors' positions on each issue elicits how much overall support there is for each issue. A strong opposition countered by a strong proponent has the same effect as two undecided positions. So the value of the sum may be less informative than how it was constituted.

Once the estimates are made for each actor, the next step is to calculate the weights each actor contributes to the decision. This is done by multiplying *position* times *salience* times *power* for each actor. After each actor's weight (*position* times *salience* times *power*) is calculated, the sum of the actors' weights is determined by adding them.

The next step is to estimate the probability of the issue (decision) to be adopted.

First, the scores of all the actors supporting the decision are added and this column total is "A." The scores of all the actors opposing the issue are added and this column total is "B." Minus signs are eliminated to create the absolute value. The non-zero scores of all the actors that have a neutral position are multiplied, and then added together in column "C." Columns A, B, and C are added across for each actor, and recorded in column D, giving the total of all the power weights in the particular issue.

Next, column A values (the strongest supporters and opponents) are added to one-half of the value of column C, the neutral actors' scores (recorded in column E). This group of actors constitutes the "swing voters," that is, the neutral actors are equally likely, in the future, to be either supporters or opponents of the issue. The best way to represent this fifty-fifty situation in the absence of other information is to include just half the neutral actors' scores with the positive weights. This total is the sum of the best estimate of the likely weights to be exerted in support of the issue.

The weights supporting the issue (E) is divided by the total weights in the system (D). The resulting fraction is the proportion of the positive weights in relation to the total weights and ranges from 0.0 (no chance of occurrence) to 1.00 (certainty of occurrence). It can be interpreted as the probability that the issue will be implemented. If the probability is between 0 percent and 39 percent the likelihood of occurrence is low; if the probability is between 40 percent and 59 percent, the likelihood of occurrence is uncertain; and if the probability is between 60 percent and 100 percent, the likelihood of occurrence is high. See the equation below.

$$\text{PRINCE Probability Equation} = \frac{\Sigma W_s}{\Sigma W_t} = \frac{\text{Sum of Total Positive Actor Scores}}{\text{Sum of Absolute Value of All Actor Scores}}$$

Step 4: Execute

The last step is to formulate and execute a strategy. On the assumption that the outcome is not a "sure thing," there are only four tools to increase the likelihood of the desired outcome: (1) Change the position of one or more actors on the issue; (2) Increase the salience of the issue for those actors who side with you, or decrease it for those who oppose you; (3) Build the power of those actors who side with you, or weaken the power of those who oppose you; (4) Make friends and win over your enemies.

NOTES

1. Jamal S. Al-Suwaidi, "Introduction," in *China, India and the United States: Competition for Energy Resources* (Abu Dhabi, United Arab Emirates: Emirates Center for Strategic Studies and Research, 2008), 5.
2. Al-Suwaidi, "Introduction," 5.

3. Philip Andrews-Speed and Roland Dannreuther, *China, Oil and Global Politics* (London: Routledge, 2011), 2.
4. Andrews-Speed and Dannreuther, *China, Oil and Global Politics*, 61.
5. Bernard D. Cole, *Oil for the Lamps of China—Beijing's 21st Century Search for Energy* (Washington, DC: Institute for National Strategic Studies, National Defense University, 2003), 15.
6. Cole, *Oil for the Lamps of China*, 15–16.
7. Andrews-Speed and Dannreuther, *China, Oil and Global Politics*, 31.
8. Andrews-Speed and Dannreuther, *China, Oil and Global Politics*, 31.
9. Andrews-Speed and Dannreuther, *China, Oil and Global Politics*, 45–46.
10. Hugh Courtney, Jane Kirkland and Patrick Viguerie, "Strategy Under Uncertainty," *Harvard Business Review* (November–December 1997): 68–71.
11. Iain Conn, "Energy Trends and Technologies for the Coming Decades" (Paper presented to the Harvard University Center for the Environment, March 8, 2006).
12. Conn, "Energy Trends and Technologies for the Coming Decades," 12.
13. Michael T. Klare, "Petroleum Anxiety and the Militarization of Energy Security," in Daniel Moran and James A. Russell, eds., *Energy Security and Global Politics: the Militarization of Resource Management* (New York: Routledge, 2009), 44.
14. Klare, "Petroleum Anxiety and the Militarization of Energy Security," 46.
15. Hilton L. Root, *Alliance Curse: How America Lost the Third World* (Washington, DC: The Brookings Institution), May 1, 2008, 4.
16. Attributed to James Burkhard of IHS Cambridge Energy Research Associates (CERA).
17. Jonathan N. Maack, "Scenario Analysis: A Tool for Task Managers," in *Social Analysis: Selected Tools and Techniques* (Washington, DC: World Bank), 2001, 62.
18. William D. Coplin and Michael K. O'Leary, *Introduction to Political Risk Analysis (Learning Packages in the Policy Sciences PS-24)* (New York: Policy Studies Associates), December 1983, 18.

2

The Competitive Dependency Scenario

BACKGROUND

The competitive dependency scenario is a future where India and Japan are competing with China for crude oil and petroleum products internationally from the same oil producing countries and China's oil consumption exceeds oil production domestically. China and India are heavily resource dependent but unlike India conventional oil production in China is nearing its peak. Unlike China and India, Japan is completely dependent on foreign oil. This represents the situation as it exists in the second decade of the twenty-first century.

Since 1993 China has imported crude oil in order to sustain economic growth and development. Moreover, in recent years, China has imported over 1,196,270,000 barrels of crude oil annually of which 80 percent or more is imported from the same oil producing states as the United States. Further, 90 percent of China's oil is imported by tankers originating from the Arabian Gulf and Sub-Saharan Africa. In addition, approximately 90 percent of China's oil tankers transit from the Indian Ocean to the South China Sea through the Strait of Malacca, a key strategic chokepoint in Southeast Asia, which potentially subjects China to supply disruptions caused by piracy, international terrorism and interdiction by U.S. naval forces.[1]

To avoid supply disruptions, China has developed diversification strategies like building oil pipelines from Russia, Central Asia, South Asia, and Southeast Asia to China and made port facility agreements with Burma and Pakistan. Moreover, China has modernized its navy and facilitated bilateral dialogues with its neighbors in South Asia and Southeast Asia regarding the deployment of Chinese warships from naval bases in Bangladesh, the Maldives, Myanmar, Pakistan, the Seychelles, and Sri Lanka, to protect the vital sea lines of communication (SLOCs) from the Strait of Hormuz and Horn of Africa to the South China Sea. Minimizing supply disruptions by ensuring the safe passage of China's oil tanker fleet transiting through the Indian Ocean region is critical to China's energy security because more than 50 percent of China's imported oil originates from the Near East and Sub-Saharan Africa.[2] Otherwise, China will not be able to sustain economic growth and development as well as maintain Communist Party rule.

This scenario consists of multiple independent, national, and events or issues of Chinese crude oil and petroleum exploration and acquisition that are currently occurring, have occurred and will probably continue or possibly will occur based on the most current literature available, and oil and petroleum energy statistics collected for the years 2006, 2007, 2008, 2009, 2010, 2011, and 2012 from FACTS Global Energy and 2012 from BP. Note that an analysis of the most recent statistics provides specific trends as to which states will matter in China's energy security in the future.

PROBABILITY OF OCCURRENCE

The overall probability of occurrence in the competitive dependency scenario is 93 percent. Regionally, the probability of occurrence is as follows: East Asia and the Pacific, 91 percent; South and Central Asia, 87 percent; the Middle East and North Africa (the Near East), 94 percent; Sub-Saharan Africa, 93 percent; Europe and Eurasia, 95 percent; and the Western Hemisphere, 94 percent. Therefore, regionally or overall, the competitive dependency scenario is a very probable scenario.

MACRO-STRATEGIC IMPLICATIONS FOR INDIA AND JAPAN

In this scenario, "China makes efforts to control its supplies from the oil wells all the way to China's coast, signing exclusive agreements with producer countries, buying or building oil exploring and trading companies, and developing strategies to defend the oil routes or to develop land-based lines of supply."[3] China's dependence on and pursuit of global energy resources causes it to be a *strategic energy resource competitor, politically, economically, and potentially (but highly remote) militarily,* to India and Japan in those countries where both nations obtain their petroleum products. It also depends on the level of U.S. political, economic, and military engagement and presence in the oil producing countries of the world.

Political competition for energy resources between China, India, and Japan is more likely to occur if each wants preferential access to scarce oil resources especially in Asia and the Persian Gulf. Ashley Tellis opined:

> Under such circumstances, Sino-Indian rivalry in Central Asia could take new forms, with both sides jockeying for preferential access in exchange for technology, economic assistance, military technologies, and possibly even guarantees.[4]

With respect to Sino-Indian competition in the Middle East, Tellis further opined:

> Sino-Indian competition over the Persian Gulf is likely to take political rather than military form.[5]

Moreover, Philip Andrews-Speed and Ronald Dannreuther stated:

> For obvious reasons, diplomatic efforts relating to energy have been focused on selected targets: those countries with the largest reserves . . . Middle East . . . Kazakhstan, Russia and Turkmenistan . . . Iran, Venezuela and Sudan . . . Angola and Nigeria . . . and countries that lie along existing or potential oil transportation routes, particularly in South Asia, for example Myanmar.[6]

However, competition is dependent on whether the United States remains engaged in world affairs especially in the Persian Gulf. Tellis opined:

> Any emerging bilateral competition with respect to the Persian Gulf would only become attenuated because of the overwhelming power and influence of the United States in the region.[7]

So, if the United States should withdraw from world affairs especially in Asia and the Near East, competition could intensify. China could decide to abandon its noninterference policy if the competition for scarce oil resources becomes great. In this respect, competition may take other forms, such as military competition. This is very unlikely because of the likelihood of U.S. intervention to ameliorate any potential crisis situation. It is also important to note that political competition may occur indirectly between China and India over Pakistan because of China's economic and security ties to Pakistan.

Economic competition derives from trade competition between China, India, and Japan in the form of energy trading blocks between the oil producing states and Chinese oil companies. If energy trading blocks favor

China's national oil companies (NOCs) over Indian and Japanese national oil companies and independent oil companies (IOCs) operating in politically non-aligned countries in the developing world, India and Japan's ability to acquire cheap oil may be undermined especially if China is able to use lucrative infrastructure investments to entice oil suppliers to favor China's national oil companies over India and Japan's national oil companies and independent oil companies in exploration and production contracts. This action could effectively shut out oil companies like India's Oil and Natural Gas Corporation (ONGC), ONGC Videsh Limited (OVL), Indian Oil Corporation Limited (IOC), Oil India Limited (OIL), Cosmo Oil, and Japan's Inpex Corporation, and JX Nippon Oil & Gas Exploration Corporation. India and Japan depend on corruption-free access to reserves so that their oil companies can compete on a level playing field.

China's pursuit of global energy resources influences the price of crude oil and petroleum globally. The U.S. Congressional Budget Office (CBO) stated in its 2006 report, *China's Growing Demand for Oil and Its Impact on U.S. Petroleum Markets*:

> World oil prices have been rising since 2004, driven in part by the increases in China's demand for crude oil and refined petroleum products. Those increases were most prominent for the light petroleum products—gasoline and diesel—that are used primarily for transportation. Over the next five years, the pace of growth in China's demand for oil in general, and for transportation fuels in particular, could be a key factor contributing to further increases in the prices for crude oil and refined petroleum products.[8]

Economic competition may be manifested in the business tactics Chinese NOCs used to secure oil deals. According to Philip Andrews-Speed and Roland Dannreuther, Chinese NOCs use corporate tactics to acquire the rights to explore for oil and gas or to develop and produce from proven fields:

> Participating in a open tender, possibly in a licensing round; buying share of a field from an owner seeking to reduce their share or to sell out completely; buying an entire field or suite of fields from a company, or buying a local subsidiary from a multinational company; buying shares in an oil company in order to gain access to assets in a specific country; forming a joint venture with a local company; negotiating directly with the host government or the NOC to gain rights to assets or acreage; and finally a whole company.[9]

Chinese NOCs have pursued every one of these tactics in their overseas expansion in such countries like Sudan, Iran, Kazakhstan, Russia, and Angola.[10] In some cases, Chinese NOCs have received direct and explicit political support from their government that may have been backed by loans from one of China's development banks, such as the China Export-Import (EXIM) Bank.[11] Such was the case in Angola, Nigeria, and Sudan.[12]

In some cases, China overbid for oil projects when the prize was substantial. Such was the case in Angola, Kazakhstan, Saudi Arabia, and Venezuela as well as in the failed bids for Slavneft and Unocal.[13] The bidding strategy is assisted by the NOCs strong financial positions. Chinese NOCs were able to receive favorable tax treatment from their government in comparison to other sectors and take advantage of low-interest free loans from state-owned policy banks.[14]

In addition to cash and loans, China has drawn on a wider array of instruments in order to gain access to reserves. According to Philip Andrews-Speed and Ronald Dannreuther:

> The contracts to explore and develop oil and gas fields may be explicitly or implicitly linked to the construction of oil refineries or export pipelines within the host country and to long-term agreements to export oil to China. In addition, Chinese oil companies may bring their own service and construction companies. Investments in Kazakhstan and Turkmenistan have also linked in production, export agreements and pipelines.[15]

China has also baited NOCs from the host country to invest in refinery, petrochemical and other downstream projects in China.[16] Saudi Aramco constructed a new refinery in Qingdao in Shandong providence in 2008 while Kuwait Petroleum Company and Qatar Petroleum Company agreed to construct a refinery and petrochemical plants.[17]

Chinese NOCs have also undertaken most of their investments in partnership with one or more other companies: Sudan, Russia, Algeria, Libya, Kazakhstan, Uzbekistan, Venezuela, Nigeria, Iran, and Myanmar.[18]

Lastly, Chinese NOCs has chosen partners NOCs from other developing countries. In Sudan, Petronas of Malaysia and ONGC of India are partners of China National Petroleum Corporation (CNPC).[19] ONGC also partnered with CNPC in Syria and China Petroleum and Chemical Corporation (Sinopec) in Colombia and Cote D'Ivoire.[20]

Based on the foregoing description, Chinese NOCs have expanded in the international oil and gas arena with increasing determination and speed.[21] Further, notwithstanding joint partnerships in the Near East, Sub-Saharan Africa, and the Western Hemisphere, it would be fair to conclude that China has engaged in economic competition for energy resources in the same regions.

India and Japan also face potential *military competition* from China over energy resources in the near future. China's defense industry and its military (People's Liberation Army [PLA]) find opportunities under this scenario to advance its foreign policy interests and improve its military relations with the oil producing nations. As the United States has often done, China can further their military relations with the oil producing nations in the wake of successful diplomatic and commercial ties, which, in turn, may have serious military consequences for India and Japan such as resorting to various military measures to protect their energy assets, not withstanding their accession to international conventions such as the Law of the Sea. These measures include: direct seizure of energy assets by military means; destruction of energy assets to deny their use to rivals like the United States, India, and Japan; military competition arising from competitive efforts to exploit energy resources either on land or sea; military protection of, or attacks upon energy production and transportation infrastructure, including oil fields, refineries, pipelines, port facilities, and so on; conveyance of major military assets to regional energy producers in exchange for preferential market; and active military control of military straits through which energy assets move.[22] Military competition may also appear in the form of an arms race between China, India, and Japan caused by the uncertainty of a Chinese naval development, the deployment of warships in the East China Sea and South China Sea, and the potential use of naval bases and port facilities in the nations that straddle the Indian Ocean such as Bangladesh, the Maldives, Myanmar, Pakistan, the Seychelles, and Sri Lanka to support Chinese naval operations such as protecting Chinese oil tankers transiting the sea lines of communication from the Persian Gulf, the Red Sea, and East Africa. Ronald O'Rourke stated that:

> In addition to preparing for a Taiwan contingency, the PLA has been developing new platforms and capabilities that will extend its operational reach to address other concerns within the East and South China Seas, and possibly to the Indian Ocean and beyond the second island chain in the western Pacific.[23]

Although China's intentions may be honorable and in the absence of transparency, Chinese naval modernization and the deployment of its warships to the East China Sea, South China Sea, and the Indian Ocean apart from anti-piracy operations in the Gulf of Aden may cause regional instability. David A. Chin opined that:

> The danger will come from the instability caused by the regional naval arms race stemming from the perceived threat of the PLAN to affect the energy security of its neighbors.[24]

Military conflict between China and India or China and Japan such as a kinetic exchange is highly remote. Should it occur, it would be unintended and undesired. However, a conflict is more likely between China and Japan than China and India because of unresolved tensions between China and Japan. Richard C. Bush offers three explanations for these tensions: "general competition for power, historical memories and interaction on specific issues" like the East China Sea dispute.[25]

It is important to note that Chinese foreign military operations are a strategic level activity in support of the larger, foreign, diplomatic, economic, and security agenda established by China's leadership.[26] David H. Shinn opined that:

> The PLA does not engage in freestanding military initiatives conducted by military professionals for military reasons. In other words, the PLA is not an independent actor; it must coordinate its activities with the party and the state bureaucracy. The PLA maintains an ambitious global program of military exchanges and training programs.[27]

Overall, China's demand for oil may shape or influence its military policies and may complicate any Indian or Japanese efforts for preserving regional stability and cooperation through military means. Much depends on whether the United States remains actively engaged politically, economically, and militarily in oil resource rich regions of the world.

As well as the macro strategic implications for India and Japan, it is essential to explore and discuss how specific geographic regions will comparatively affect India and Japan politically, economically, and militarily because the China, India, and Japan have varied levels of political, diplomatic, economic, and military relations with and influence over energy suppliers. Further, any scholarly work on global energy security, especially where India and Japan may be affected by Chinese political, economic, or military relations or influence, would not be credible without appreciating the regional differences. Therefore, regional analyses are made in this energy scenario for a greater understanding of how each region specifically affects Indian and Japanese oil energy security.

REGIONAL ANALYSES

In the competitive dependency scenario, regional analyses of China's global oil trade and its implications for India and Japan are important in order to determine how their oil energy security is affected. They are the central focus of PRINCE analysis because as a risk analysis assessment method, the PRINCE model quantitatively determines whether China, India, and Japan will peacefully share the world's oil reserves or whether competition for crude oil and petroleum reserves will lead to conflict.

Realistically, the impact of China's global oil trade on India and Japan will vary (positively, negatively, or none) based on the extent of overall Chinese, Indian, and Japanese diplomatic/political, economic/trade, and military/security relations, aid and influence the amount of oil energy resources available in each region.

In the competitive dependency scenario, China's impact on the global energy trade is best understood through its strategic partnerships. Dawn Murphy opined that:

> Starting in the mid-1990s, China introduced a new diplomatic mechanism called "strategic partnership." These strategic partnerships are established with individual countries and groupings of countries. China's strategic partnerships are not military alliances (or quasi-military alliances).[28]

Evan S. Medeiros further opined on strategic partnerships stating:

> For China, *strategic partnership* has a different meaning from the Western connotation of the term. A partnership is strategic for two reasons: (1) It is comprehensive, including all aspects of bilateral relations (e.g. economic, cultural, political, and security), and (2) both countries agree to make a long-term commitment to bilateral relations, in which bilateral problems are evaluated in that context and, importantly, occasional tensions due not derail them.[29]

It would be beneficial to study India and Japan's strategic partnerships in the Chinese sense. Therefore, China, India, and Japan will be examined comparatively in terms of their strategic partnerships by individual countries and grouping of countries.

China's cultural relations will not be examined because comparative data is not available from India and Japan. Nevertheless, China has been very active in promoting the Chinese language and culture. It has established numerous Confucius Institutes in key countries that serve Chinese strategic interests: Azerbaijan, Bangladesh, Egypt, Iran, Myanmar, Nigeria, Sri Lanka, Sudan, Seychelles, and the United Arab Emirates.[30] The institutes seek to improve China's image abroad and assuage concerns about "a China threat" in the context of the country's increasingly powerful economy and military.[31] Many scholars have characterized the institutes as an exercise in adroit Chinese soft power,[32] while others have suggested that the institutes are engaged in intelligence collection.[33]

In terms of available hydrocarbon resources, proven crude oil reserves as of October 5, 2013 are shown according to the U.S. Energy Information Administration (EIA). See table 2.1.

The Near East holds the majority of proven oil reserves and is clearly the most important region of the world in terms of energy security for China, India, and Japan. This book focuses not only on the Near East and the Western Hemisphere but also Russia, the Asia-Pacific, Central Eurasia, Sub-Saharan Africa, and Western Europe where Chinese, Indian, and Japanese oil interests overlap. Key oil producing states such as Angola, Canada, Iran, Nigeria, Saudi Arabia, Sudan, and Venezuela are examined.

Table 2.1 Proven Oil Reserves

Region	Billion Barrels	% of World Total
Russian Federation	80	5
Asia-Pacific	47.2	3
Europe and Eurasia	50.9	3
Near East (Middle East & North Africa)	867	54
Africa (Sub-Saharan Africa)	62.5	4
Western Hemisphere (North & South America)	509.2	31
Total	1,616.8	100

Source: U.S. Energy Information Administration, 2013.

Chapter 3 examines East Asia and the Pacific and explores China's territorial claims in the East China Sea and South China Sea.

Chapter 4 looks at South and Central Asia including the oil pipelines into China and China's geopolitical ambitions in the Indian Ocean.

Chapter 5 focuses on the Middle East and North Africa including China's dangerous relationship with Iran.

Chapter 6 examines Sub-Saharan Africa including China's emerging strategic partnerships in the region including Angola, Nigeria, and Sudan.

Chapter 7 describes China's oil energy relationships in Europe and Eurasia including Russia, Norway, and the United Kingdom. China's potential oil exploration in the Kara Sea is also discussed.

Chapter 8 analyzes the Western Hemisphere emphasizing China's burgeoning economic relationships with Venezuela.

Chapter 9 concludes the discussion and summarizes the key findings.

NOTES

1. Flynt Leverett, "Resource Merchantism and the Militarization of Resource Management: Rising Asia and the Future of American Primacy," in Daniel Moran and James A. Russell, eds., *Energy Security and Global Politics: the Militarization of Resource Management (*New York: Routledge, 2009), 217.
2. Jon. B. Alterman and John W. Garver. *The Vital Triangle: China, The United States and the Middle East* (Washington, DC: CSIS Press, 2008), 6.

3. Volker Stanzel, "The EU and China in the global system," in David Shambaugh, Eberhard Sandschneider, and Zhou Hong, eds., *China–Europe Relations: Perceptions, Policies and Prospects* (New York: Routledge, 2008), 258.

4. Ashley J.Tellis, "China and India in Asia," in Francine R. Frankel and Harry Harding, eds., *The India–China Relationship: What the United States Needs to Know* (New York: Columbia University Press, 2004), 158.

5. Tellis, "China and India in Asia," 161.

6. Philip Andrews-Speed and Roland Dannreuther, *China, Oil and Global Politics* (London: Routledge, 2011), 84.

7. Tellis, "China and India in Asia," 161.

8. U.S. House, Congressional Budget Office, *China's Growing Demand for Oil and Its Impact on U.S. Petroleum Markets,* Pub. 2561, April 2006, 1.

9. Andrews-Speed and Dannreuther, *China, Oil and Global Politics,* 80.

10. Andrews-Speed and Dannreuther, *China, Oil and Global Politics,* 80.

11. Andrews-Speed and Dannreuther, *China, Oil and Global Politics,* 80.

12. Andrews-Speed and Dannreuther, *China, Oil and Global Politics,* 80–81.

13. Andrews-Speed and Dannreuther, *China, Oil and Global Politics,* 81.

14. Andrews-Speed and Dannreuther, *China, Oil and Global Politics,* 81.

15. Andrews-Speed and Dannreuther, *China, Oil and Global Politics,* 81.

16. Andrews-Speed and Dannreuther, *China, Oil and Global Politics,* 81.

17. Andrews-Speed and Dannreuther, *China, Oil and Global Politics,* 81–82.

18. Andrews-Speed and Dannreuther, *China, Oil and Global Politics,* 82.

19. Andrews-Speed and Dannreuther, *China, Oil and Global Politics,* 83.

20. Andrews-Speed and Dannreuther, *China, Oil and Global Politics,* 83.

21. Andrews-Speed and Dannreuther, *China, Oil and Global Politics,* 83.

22. Daniel Moran and James A. Russell, "Introduction: the militarization of energy security," in Daniel Moran and James A. Russell, eds., *Energy Security and Global Politics: the Militarization of Resource Management* (New York: Routledge, 2009), 7.

23. Ronald O'Rourke, *China Naval Modernization: Implications for U.S. Naval Capabilities—Background and Issues for Congress.* CRS Report RL33153, (Washington, DC: the Library of Congress, December 1, 2010), 5.

24. David A. Chin, "The Prospect of China's Access to Naval Facilities in Burma and the Ramifications for Regional Stability" (master's thesis, Naval Postgraduate School, 2007), 1.

25. Richard C. Bush, *The Perils of Proximity: China–Japan Security Relations* (Washington, DC: Brookings Institution Press, 2010), 22.

26. David H. Shinn, "Military and Security Relations: China, Africa, and the Rest of the World" in Robert I. Rotberg, ed., *China Into Africa: Trade, Aid, and Influence* (Washington, DC: Brookings Institution Press, 2008), 162.

27. Shinn, "Military and Security Relations," 162.

28. Dawn Murphy, "China and the Middle East," testimony, June 6, 2013, before U.S.–China Economic and Security Review Commission.

29. Evan S. Medeiros, *China's International Behavior: Activism, Opportunism, and Diversification* (Santa Monica, CA: RAND Corporation, 2009), 82.

30. Confucius Institute Online, "Worldwide Confucius Institutes," accessed January 17, 2014, http://www.chinesecio.com/m/cio_wci.

31. Xiaolin Gao, *Repacking Confucius: PRC Public Diplomacy and the Rise of Soft Power* (Stockholm, Sweden: Institute for Security and Development Policy), January 2008, 26.

32. Peter Schmidt, "At U.S. Colleges, Chinese-Financed Centers Prompt Worries About Academic Freedom," *The Chronicle of Higher Education*, October 22, 2010, A8, accessed January 15, 2014, http://chronicle.texterity.com/chronicle/20101022a?pg=8#pg8.

33. Fabrice De Pierrebourg and Michel Juneau-Katsuya, *Nest of Spies: The Startling Truth About Foreign Agents At Work Within Canada's Borders* (Canada: HarperCollins, 2009), 160–162.

3

East Asia and the Pacific

This region is comprised of East Asia and Southeast Asia including Oceania (Australia). Even though this region is vital for trade and investment, it is not crucial in terms of oil trading for China, India, and Japan. Crude and refined petroleum imports from this region are relatively low compared to the other resource-rich regions of the world.

Similarly, the region is becoming less significant for China in terms of hydrocarbon energy trading. China has imported less crude oil and petroleum from East Asia and the Pacific than any other region except Europe from 2006 to 2012. Many oil exporting nations, especially those in Southeast Asia, are gradually approaching a decline in their hydrocarbon energy reserves. Thus, the region is not really a "battleground" region for Sino-Indian or Sino-Japanese oil competition except in the East China Sea and South China Sea.

Although China imports less crude oil and petroleum from East Asia and the Pacific than any other region in the world, the region is not any less important in terms of energy security. Approximately 93.5 percent of China's crude oil and petroleum imports are transported by sea using ocean-going tankers and transit through either the Strait of Hormuz or Strait of Malacca, two major strategic chokepoints that are constantly patrolled by the United States Navy.

Because the PLA Navy has neither the military nor logistical capability to patrol these vital areas and challenge U.S. naval supremacy, China has to rely on means other than military power to ensure the steady and uninterrupted flow of oil from oil rich regions of the world.

One way for Beijing to ensure the safe transit of its foreign oil was to become more regionally involved through trade, political dialogue, and cultural exchanges. While China appears to be genuinely benevolent, most nations are wary of its true intentions for East Asia and the Pacific region. Yet, many East Asian and Pacific nations have engaged China on various levels, which has enhanced their relationship with the Middle Kingdom. For the most part, China's soft power approach (through cultural and economic means) or so-called "charm offensive" has paid enormous dividends in recent years resulting in stronger political, economic, and cultural ties. However, such "ties" bear closer scrutiny because of variations in actual political, economic, and military relations between China and the other nations of East Asia and the Pacific region.

EAST ASIA

The PRINCE model predicts that China will continue to explore for oil domestically by 91 percent, in the littorals by 98 percent, in the East China Sea by 71 percent and will continue to import oil from Mongolia and South Korea by 98 percent and 98 percent respectively.[1]

Political, Economic and Military Implications for India and Japan

China's hydrocarbon energy trading with East Asia (Mongolia and South Korea) is unlikely to have political, economic, or military ramifications for India and Japan because India and Japan do not import crude or refined petroleum from either Mongolia or South Korea. Moreover, India and Japan have diplomatic and economic ties with both Mongolia and South Korea according to their foreign ministries as well as cultural ties. In terms of military relations, India and Mongolia are friendly and cordial and are signified by high-level visits, training cooperation, and joint exercises.[2] India and Mongolia held a joint working group on defense cooperation in May 2011, signed an agreement on defense cooperation in July 2011, and participated in the joint military exercise *Nomadic Elephant* in September 2011.[3] The Indian Army Chief of Staff also visited Mongolia in September 2011 and participated in national celebrations honoring the Mongolian armed forces.[4]

Indo-South Korean military relations are unknown. India provided United Nations (UN) peacekeeping forces to Korea from 1953 to 1954. India and South Korea do not appear to have any bilateral military agreements, defense cooperation or military exchanges.

Japan and Mongolia signed a memorandum of understanding (MOU) on defense cooperation and exchange in January 2012 and conducted an exchange visit between their respective vice-ministers of defense on capacity building and regional issues in October 2012 according to the Defense of Japan.[5] In June 2011, Japan and Mongolia participated in the Global Peace Operations Initiation Capstone training exercise (*Ayala Guardian 11*) with the United States, Bangladesh, Indonesia, Malaysia, Nepal, the Philippines, Thailand, and Vietnam.[6] Further, from July 2011 to August 2011, Japan and Mongolia participated in the multilateral joint training exercise *Conquest 11*.[7]

Japanese–South Korean military relations are very close and are signified by high-level talks between the heads of state and defense, regular discussions between defense military representatives, unit-level exchanges, and trilateral defense cooperation between Japan, South Korea, and the United States.[8] Both countries also share many strategic interests and basic values as allies of the United States, such as permitting the stationing of U.S. armed forces.[9]

The Political, Economic, and Military Implications of China's Domestic Hydrocarbon Operations

Domestic hydrocarbon operations (exploration, discovery, extraction, and refinement) in territorial China such as the Tarim Basin in Xinjiang, Bohai Bay, and coastal China are unlikely to have any negative diplomatic, economic, or military consequences for India and Japan such as political, economic, or military competition for strategic energy resources. These operations are sovereign in nature and do not involve India, Japan, or any other country except in a capacity where technologically advanced countries like the United States may be able to provide public and private technical expertise to assist China in acquiring its natural hydrocarbon energy resources from underground and below the ocean floor. If China conducts domestic hydrocarbon operations and seeks greater international cooperation from the United States through public and private technical assistance, then both countries will benefit on many levels such as increasing their strategic petroleum reserves (SPR) as well as lowering global oil prices. Therefore, it behooves both China and the United States to engage each other and work constructively to help reduce China's dependency on foreign hydrocarbons. This will benefit the East Asian and Pacific region, especially India and Japan, which greatly depend on imported crude and refined petroleum to sustain their economies and militaries.

China–Japan Flashpoint: The East China Sea Dispute (Diaoyu/Senkaku Islands)

The East China Sea dispute has been a major point of contention between China and Japan since June 1971 when the United States transferred the administration of the Diaoyu/Senkaku Islands (DSI) to Japan under the Okinawa Reversion Treaty. Previously, Japan controlled the DSI from 1895 until its surrender at the end of World War II. Conversely, China claimed the discovery and control of the DSI from the fourteenth century. Although Japan occupies the DSI, little attention was given towards DSI sovereignty until 1969 when a UN report revealed that large hydrocarbon deposits exist in its surrounding waters.

Basically, the dispute over natural resources in the East China Sea revolves around two issues: demarcating the sea boundary between each country and sovereignty over the Daioyu/Senkaku Islands. Both countries claim exclusive sovereignty of eight uninhabited islands (five uninhabited islets and three barren rocks) presumably for strategic and political reasons, as claims of ownership are used to bolster claims to the surrounding sea and its natural resources. Taiwan also claims the DSI as sovereign territory but has not actively pursued its natural resources. See table 3.1.

In terms of natural resources, the contested undersea beds are estimated to hold 17.9 million barrels of crude petroleum and 363.3 billion cubic feet of natural gas. Further, the entire East China Sea region is estimated to hold unproven reserves of approximately 100 billion barrels of crude petroleum and 7 trillion cubic feet of natural gas making it a major source of political tension between China and Japan. Japan's interest reflects the fact that it lacks natural resources and is heavily dependent upon imported oil and gas to sustain its national economy and military.

Since the discovery of oil and gas, efforts to mediate the dispute peacefully have been futile. Both China and Japan have held at least eleven

Table 3.1 Daioyu/Senkaku Islands

Japanese name	Chinese name
Uotsuri-jima	Diaoyu Dao
Taishō-jima	Chiwei Yu
Kuba-jima	Huangwei Yu
Kita-Ko-jima	Bei Xiaodao
Minami-Ko-jima	Nan Xiaodao
Oki-no-Kita-iwa	Da Bei Xiaodao
Oki-no-Minami-iwa	Da Nan Xiaodao
Tobise	Fei Jiao Yan

official talks since October 2004 to no avail. Bilateral offers for joint development of resources were made but not agreed to. Finally, after years of stalemate, Japan purchased and nationalized three (Minami-Ko-jima/Nan Xiaodao, Kita-Ko-jima/Bei Xiaodao and Uotsuri-jima/Diaoyu Dao) of the eight uninhabited islands from the Kurihara family for 2.05 billion yen ($26.2 million) on September 11, 2012.

Nationalization of the islands has reignited political tensions between China and Japan setting off a series of threats, protests, and demonstrations in both Asian countries as well as increased civil and coast guard maritime patrols around the disputed island group. It is uncertain whether military conflict will breakout between China and Japan but China must be able to save face without having to resort to force.

The United States does not wish to take sides in this issue but has offered to mediate because it does not wish to see any instability in the Asia-Pacific especially in the Western Pacific because of the potential disruptions to international trade. Both China and Japan would stand to lose economically if the dispute leads to a breakdown in relations that interrupts production and triggers boycotts. Bilateral trade has tripled over the past decade to more than $340 billion. China is now Japan's major export market and Japanese investment in China has been running at twice that of the United States and South Korea in recent years.[10]

SOUTHEAST ASIA INCLUDING OCEANIA

The PRINCE model predicts that China will continue to explore for oil in, and import oil from, Southeast Asia based on a probability of 85 percent or greater. However, it is uncertain as to whether China will unilaterally explore, discover, extract, and refine crude oil from the Spratly Islands based on a probability of 58 percent.[11]

Political, Economic, and Military Implications for India and Japan

China imports crude oil from Australia, Brunei, Burma (Myanmar), Indonesia, Malaysia, New Zealand, Papua New Guinea, the Philippines, Thailand, and Vietnam with the largest oil suppliers being Australia (over 1 percent annually) and Malaysia (less than 1 percent annually) according to FACTS Global Energy. In comparison, India imports crude oil from Australia (less than 1 percent), Brunei (less than 1 percent), Malaysia (over 1 percent) and Japan imports crude oil from Australia, Brunei, Indonesia (approximately 4 percent annually), Malaysia, Papua New Guinea, and Vietnam according to FACTS Global Energy. China, India, and Japan also have oil exploration and production operations in the region. See appendices B, C, and D. China's hydrocarbon energy trading in Southeast Asia and Oceania is unlikely to result in political, economic, or military competition for crude oil because of the strong diplomatic and economic ties that India and Japan enjoy in the region. Further, many Southeast Asian nations see China as a hegemon which is trying to dominate the region politically, economically, and militarily. In this respect, Southeast Asian nations deeply value the U.S. security umbrella against a potentially expansive China despite widespread improvements in public perceptions of China and parallel declines in perceptions of the United States.[12] Thus, the United States "has the added advantage of being seen by Southeast Asian and other Asian government leaders as Asia's 'least distrusted power' as these leaders seek to enhance their nationalistic ambitions and independence of action among other Asian governments, including and especially China, that have territorial and other ambitions directly at odds with their national interests."[13] Moreover, the strategic importance of Southeast Asia as the site of future oil supplies is low. "Southeast Asia simply does not have the potential for future oil supplies to be a major part of the solution to China's security angst,"[14] causing it to remain a sideshow at least insofar as oil supplies are concerned. However, Southeast Asia will maintain its strategic importance to China because of the regional sea lines of communication (SLOCs).

"China's growing influence derives mainly from its role as a market for the region's natural resources," even though the region contains approximately 3.3% of the world's proven oil reserves, while the "United States maintains its influence based on its military presence, foreign direct investment, its market for the region's manufactured goods, military and development assistance and educational opportunities."[15]

However, China's influence and hydrocarbon energy security in Southeast Asia could be set back by disruptions in the sea lines of communication such as: a sovereignty dispute in the South China Sea which threatens military conflict; terrorism and international crime, usually falling under the rubric of "piracy" in the Malacca Strait;[16] and a U.S. navy blockade in the Malacca Strait in the event that China militarily invades Taiwan. Since 60 percent of China's crude oil imports currently pass through the Malacca Strait and the South China Sea with an estimated increase towards 75 percent by the year 2015,[17] the safety and security of these sea lines of communication becomes a critical factor not only to China's regional influence and hydrocarbon energy security, but also its economic growth and continued rule by the Chinese Communist Party because "the threat of economic stagnation (caused by maritime trade disruptions) raises real risks of social instability, which could in turn threaten the continued political monopoly of the Chinese Communist

Party."[18] Since oil imports destined for China from Africa and the Arabian Gulf are vulnerable to terrorism, piracy, U.S. naval interdiction in and around the Indian Ocean and the Strait of Malacca and potential military conflict in the South China Sea, it behooves China to pursue various hydrocarbon energy security strategies that will reduce its perceived vulnerabilities associated with its maritime oil imports.

China is considering, among various hydrocarbon energy security strategies, one that is based on the diversification of energy import routes through the development of Pakistani and Burmese transnational pipelines. Beijing is also considering the idea of building a shipping canal through Thailand's Kra Isthmus to divert her maritime fleet directly into the Gulf of Thailand. As a direct consequence of the growing volume of crude oil flowing to China by tanker through the Indian Ocean and South China Sea, China is developing its naval capability to reach beyond coastal waters to the open ocean or the blue-water, whereby naval assets will be used to protect its global maritime shipping interests (including the oil tankers) transiting through Indian Ocean and elsewhere and will include large capital ships like a nuclear aircraft carrier (CVN) and Russian made diesel submarines.[19] This notion is known as the "militarization of resource management" resulting from energy insecurity.

Further, in tandem with the resource militarization Herberg opined,

> China is developing naval port access agreements along the coasts of the Indian Ocean, most importantly in Pakistan, Myanmar (Burma) and Bangladesh, that would allow its ships and submarines to operate more effectively there and into the Strait of Malacca.[20]

This strategy is appropriately named the "string of pearls strategy" by a Pentagon report. "The 'String of Pearls' describes the manifestation of China's rising geopolitical influence through efforts to increase access to ports and airfields, develop special diplomatic relationships, and modernize military forces that extend from the South China Sea through the Strait of Malacca, across the Indian Ocean, and on to the Arabian Gulf."[21]

In addition, Herberg opined:

> China is rapidly accelerating its own submarine development along with the acquisition of sophisticated Russian-made submarines capable of operating long distances.[22]

China is already paving the way for this new submarine force by expanding its submarine basing facilities on Hainan Island in the South China Sea.

Finally, China has embarked on a program to build its own tanker fleet to move China's oil supplies because only 10 percent of China's oil imports arrive by Chinese tankers.[23] It should be noted that strong diplomatic relationships are a must with the maritime nations of Indonesia, Malaysia, the Philippines, Singapore, and Thailand because it is critical to China's energy security strategy in Southeast Asia, especially since these countries are geographically situated along major maritime trade routes.

The "'militarization' of resource management describes a growing inclination of energy producers and consumers to base decisions about their behavior in the global energy markets on political and strategic calculations as well as commercial and economic considerations."[24] In China's case, resource militarization evolves from a mercantilist energy policy that reflects its sense of weakness and vulnerability regarding reliable access to energy supplies.[25] This sense of weakness provides the rationale for direct state intervention and support from Beijing rather than plenipotentiary oversight by the three Chinese national oil companies (CNPC, CNOOC and Sinopec). Further, Chinese energy policy, which provides the basis for resource militarization, is shaped by three factors which are mercantilist in nature: a general mistrust of global energy markets; that distrust of the energy markets is aggravated by the perception that the energy markets are controlled by the United States; and that in terms of energy sector capabilities, Beijing feels that China is working from a position of weakness and must play "catch-up."[26] Thus, China is an outstanding example of both mercantilism and resource militarization because its "rapidly growing navy is universally viewed as derivative of that nation's export-driven economic expansion, and corresponding dependence on raw materials imported by sea."[27] China's naval capability is becoming more Mahanian (global and strategic) in nature, growing in tandem with its sea-borne dependent economy and providing protection to its merchant fleet on the high seas while supplementing its continental strategy of energy import route diversification through Burma and Pakistan.

Beyond the goal of establishing a sphere of economic influence in Southeast Asia that is potentially inimical to India and Japan, China seeks to ensure its energy security to sustain economic growth and the longevity of the Chinese Communist Party. Mikkal E. Herberg opined that:

> As China's economy grows and its energy needs boom, energy supplies and transit are likely to become increasingly important factors in China's relations with Southeast Asia and the region's key players. In other words, the geopolitics of energy in the region is likely to become an important force in shaping overall regional relations.[28]

Herberg's comment raises questions as to how China's hydrocarbon energy trading with the countries of Southeast Asia have shaped economic relations vis-à-vis India and Japan in such a way that it has strategic implications of an economic nature. A more in-depth analysis of China's hydrocarbon energy trading with Southeast Asia is required to see what effect, if any, it has on India and Japan economically.

Michael Chambers believes that China must develop greater economic capacity and be viewed as generating a greater positive political presence within the region, without its military appearing to be threatening, if it is to emerge as a great power within Asia.[29]

China–Japan Flashpoint: The South China Sea Dispute

Six Asian nations—China, Brunei, Malaysia, the Philippines, Taiwan, and Vietnam—claim ownership of all or part of the Spratly Islands, which are used to bolster claims to the surrounding sea and its natural resources. In terms of natural resources, the U.S. Geological Survey estimates that the South China Sea has 28 billion barrels of crude petroleum and 266 trillion cubic feet of natural gas. Thus, the region holds great strategic value for not only securing the SLOCs but also the recourse beneath the ocean floor. Admiral Patrick Walsh USN opined the South China Sea—which is heavily traveled by shipping companies, including tankers transporting oil from the Persian Gulf to East Asian nations—is vital to the Asia-Pacific region.[30]

China's claim covers three-fourths of the total area of the South China Sea. It originated with the Kuomintang government of the Republic of China in 1947 and was modified from an "eleven-dotted" line to a "nine-dotted" line by the Chinese Communist Party after the People's Republic of China was formed in 1949. The legacy of the nine-dotted line is seen by some Chinese government officials and the Chinese military as providing historical support for Chinese claims over the region.[31]

In 2012 China was very active in asserting absolute sovereignty in the region. China engaged in a naval standoff with the Philippine navy over the Scarborough Shoal establishing the Sansha garrison on Yongxing (Woody) Island in the Paracel Islands. In addition, China launched a program to increase the number of UAVs monitoring the Scarborough Shoal, Paracel Islands, and Spratly Islands.[32]

China's unilateral exploration, discovery, extraction, and refinement of crude oil in the Spratly Islands may make the South China Sea politically unstable and disrupt the flow of international shipping traffic (including the transit of oil tankers) in the region even though China, India, and Japan import very little oil from the countries encompassing the South China Sea. This is due in part to greater Chinese military presence in the South China Sea to protect its claimed oil interests in the Spratly Islands. Second, there is an increased probability of armed conflict between China, Brunei, Malaysia, the Philippines, Taiwan, and Vietnam over conflicting sovereignty claims in the Paracel and Spratly Islands. This would most likely lead to a severe economic downturn, increasing the probability of political instability and the specter of war in the South China Sea from the disruption of crude oil and petroleum supplies.

While less than 4 percent of Japan's oil is imported from Southeast Asia according to FACTS Global Energy, it depends on unfettered access to Persian Gulf oil to sustain its economy and military. A disruption in the sea lines of communication in the South China Sea would gravely damage Japan's economy especially since it is wholly dependent on natural resources.

Although the United States is not a major importer of Southeast Asian oil, any oil supply disruptions to Japan would surely invite significant United States military intervention in the region. Thus, it behooves China to peacefully resolve any territorial disputes with its regional neighbors despite its historical claims to ownership of the South China Sea region. In this respect, the United Nations Convention on the Law of the Sea (UNCLOS) has not yet resolved the ownership disputes in the South China Sea and the region lacks commonly agreed standards for dispute settlement.

NOTES

1. Prince and Panda website, "PRINCE_SCORES.pdf," accessed October 17, 2013, http://www.princeandpanda.com.
2. Ministry of Defence, Government of India, *Annual Report 2011–2012*, 2012, 196.
3. Ministry of Defence, Government of India, *Annual Report 2011–2012*, 2012, 196.
4. Ministry of Defence, Government of India, *Annual Report 2011–2012*, 2012, 196.
5. Japanese Ministry of Defense, *Defense of Japan 2012*, 2012, 297.
6. Japanese Ministry of Defense, *Defense of Japan 2012*, 2012, 277.
7. Japanese Ministry of Defense, *Defense of Japan 2012*, 2012, 277.
8. Japanese Ministry of Defense, *Defense of Japan 2012*, 2012, 284.
9. Japanese Ministry of Defense, *Defense of Japan 2012*, 2012, 283.
10. Yukon Huang, "The Economics Behind the China–Japan Dispute," *Financial Times*, September 24, 2012, accessed October 8, 2013, http://carnegieendowment.org/2012/ 09/24/economics-behind-china-japan-dispute/dwc6.
11. Prince and Panda website, "PRINCE_SCORES.pdf," accessed October 17, 2013, http://www.princeandpanda.com.
12. Christopher M. Blanchard, et. al, *Comparing Global Influence: China's and U.S. Diplomacy, Foreign Aid, Trade, and Investment in the Developing World.* CRS Report RL34620 (Washington, DC: Library of Congress, August 15, 2008), 77.
13. Robert G. Sutter, "China's Rise, Southeast Asia, and the United States: Is A China-Centered Order Marginalizing the United States," in Evelyn Goh and

Sheldon W. Simon, eds., *China, United States, and Southeast Asia: Contending Perspectives on Politics, Security, and Economics* (New York: Routledge, 2008), 92.

14. Mikkal Herberg, "China's Search for Energy Security: the Implications for the Southeast Asia," in Evelyn Goh and Sheldon W. Simon, eds., *China, United States, and Southeast Asia: Contending Perspectives on Politics, Security, and Economics* (New York: Routledge, 2008), 79.

15. Blanchard, et. al, *Comparing Global Influence*, 77.

16. Bernard D. Cole, *Sea Lanes and Pipelines: Energy Security in Asia* (Westport, CT: Praeger Security International, 2008), 83.

17. Corey S. Johnston, "Transnational Pipelines and Naval Expansion: Examining China's Oil Insecurities in the Indian Ocean" (master's thesis, Naval Postgraduate School, 2008), 1.

18. Kenneth Lieberthal and Mikkal Herberg, *China's Search for Energy Security: Implications for U.S. Policy* (Seattle: National Bureau of Asian Research), Volume 17, Number 1, April 2006, 11.

19. *The Emergence of China Throughout Asia: Security and Economic Consequences for the United States,* Hearing before the Committee on Foreign Relations, U.S. Senate, 109th Congress, June 7, 2005, 44–45 (Statement of Mike Herberg, Director, Asian Energy Security Program, The National Bureau of Asian Research, Seattle, WA).

20. Herberg, "China's Search for Energy Security," 82.

21. Christopher J. Pehrson, *String of Pearls: Meeting the Challenge of China's Rising Power Across the Asian Littoral.* Carlisle, PA: Strategic Studies Institute, July 2006, v.

22. Herberg, "China's Search for Energy Security," 79.

23. Herberg, "China's Search for Energy Security," 77.

24. Flynt Leverett, "Resource Merchantism and the Militarization of Resource Management: Rising Asia and the Future of American Primacy," in Daniel Moran and James A. Russell, eds., *Energy Security and Global Politics: the Militarization of Resource Management* (New York: Routledge, 2009), 211.

25. Lieberthal and Herberg, *China's Search for Energy Security,* 14.

26. Lieberthal and Herberg, *China's Search for Energy Security*, 15–16.

27. See footnote 3 of Moran, Daniel and James A. Russell, "Introduction: the militarization of energy security," in Daniel Moran and James A. Russell, eds., *Energy Security and Global Politics: the Militarization of Resource Management.* (New York: Routledge, 2009), 17.

28. Herberg, "China's Search for Energy Security," 71.

29. Michael Chambers, "China's Military Rise to Great Power Status: Its Implications for the United States in Southeast Asia," in Evelyn Goh and Sheldon W. Simon, eds., *China, the United States, and Southeast Asia: Contending Perspectives on Politics, Security, and Economics* (New York: Routledge, 2008), 168.

30. Audrey McAvoy, "Adm.: Spratly Row Threatens Security, Stability," *Associated Press,* January 18, 2012.

31. Kathrin Hille, "Chinese boats fish in dangerous waters," *Financial Times,* April 24, 2012.

32. J. Michael Cole, "China increases territorial UAVs," *Taipei Times,* September 24, 2012.

4

South and Central Asia

In terms of the oil producing states, the South and Central Asian region is comprised of Kazakhstan, Turkmenistan and Uzbekistan. Pakistan does not produce oil for China, India, or Japan. Central Asia is not vital to either Indian or Japanese energy security. Thus, it is inconsequential.

In terms of energy trading, the region does not export crude or refined petroleum to either India or Japan with the exception of Kazakhstan. During fiscal year 2009, Kazakhstan exported 3 million barrels of crude petroleum to India. Further, an agreement between ONGC Videsh Limited and Kazmunaygaz on the purchase of 25 percent stake in the Satpayev Oil Block in the Caspian Sea was finalized in Astana on April 16, 2011, during the visit of Indian Prime Minister Manmohan Singh.[1] In terms of Japanese-Kazakhstani energy agreements, Japan's INPEX North Caspian Sea, Ltd. acquired a working interest in the Offshore North Caspian Sea block (Kashagan Oilfield) inside Kazakhstan's territorial waters in August 1998.[2]

The PRINCE model predicts that China will continue to explore for oil in, and import oil from, Central Asia based on the following probabilities: 95 percent (Kazakhstan), 93 percent (Turkmenistan) and 91 percent (Uzbekistan).[3]

CENTRAL ASIA

China has four sets of interests and policies in Central Asia.[4] First, China has strategic and diplomatic goals that play into China's foreign policy with respect to Sino-U.S. relations and Sino-Russian relations. Second, China seeks to safeguard the stability and security of its borders with the Central Asian nations. Third, China seeks to prevent any external influence in its western region, including support for the Uighur separatists in Xinjiang. Finally, China has energy and trade interests, which include the development of energy resources including oil and gas.

Political, Economic and Military Implications for India and Japan

None of China's interests and policies appear to challenge Indian or Japanese interests in the region because they share similar goals such as combating terrorist activity especially in the wake of the September 11, 2001, terrorist attacks against the United States.[5] Further, the region's geographic position—landlocked and remote for Indian or Japanese interests, which makes it of little strategic consequence. Moreover, in terms of Indian energy security, the region is a very minor oil producer. In addition, the region's proven oil reserves are very low—4 percent—compared to other resource rich regions of the world like Africa, the Near East, and the Western Hemisphere. Finally, China is just beginning to realize the region's importance to its energy security especially for diversification purposes. Therefore, it is unlikely that China's energy trading with the region will result in Sino-Indian or Sino-Japanese diplomatic competition for oil energy resources.

China's oil energy stakes with Central Asia is unlikely to result in either Sino-Indian or Sino-Japanese political, economic, or military competition for oil resources because India and Japan do not have significant oil interests in the region. However, if China was granted exclusive access to the region's oil resources, then rivalry could take on a new dynamic. Under such circumstances, dyadic or even triadic rivalry could result in each country jockeying for preferential access to oil resources in exchange for technology, economic assistance, military technologies, and possibly even security guarantees.[6] However, the region is strategically vital to China and India's national security because it is a potential area of political instability including a haven for terrorists to recruit, train, and operate from.

SOUTH ASIA AND THE INDIAN OCEAN

South Asia does not produce oil for China, India, or Japan. However, South Asia is critical to their energy security because the region straddles the Indian Ocean and contains vital shipping lanes by which crude oil is transported by oil tankers to China, India, and Japan from the Persian Gulf and Africa. The scarcity of energy resources in Japan makes the

SLOCs that run from the Persian Gulf, the Indian Ocean, the Strait of Malacca, the South China Sea, the Bashi Channel, and the east coast of Taiwan to Japan's vicinity and the stability of the coastal countries around the SLOCs are of crucial importance for Japan. It is of particular strategic importance to China because 25 percent of China's oil is imported from Africa and approximately 54 percent is imported from the Persian Gulf region. Since China does not have the blue-water naval capability to protect its tanker fleet beyond the South China Sea, the responsibility lies with the United States Navy to patrol the sea lines of communication (SLOCs) not only in the Persian Gulf and the Indian Ocean but also the strategic chokepoints at the eastern and western approaches to the Indian Ocean at the Strait of Malacca, the Strait of Bab-el-Mandeb (between the Red Sea and the Gulf of Aden) and the Strait of Hormuz.

Military reliance on the United States Pacific Fleet to keep the sea lines of communication open from Africa, the Red Sea and Persian Gulf to the South China Sea via the Indian Ocean poses a strategic quandary for China because China's oil lifelines could be severed in the event of an East Asian crisis such as an attack on Taiwan. In this respect, China would be powerless to stop a U.S. naval blockade of any oil tankers bound for China that were transiting through the Strait of Hormuz or Strait of Malacca.

China will also have to contend with Indian naval power in the Indian Ocean region because India uses the same sea lines of communication; relying on these sea lanes for transporting 90 percent of its oil and over 90 percent of its foreign trade; competes globally with China for oil energy resources; and has the most capable navy in the region. Further, to a lesser extent, China will have to compete with Japanese naval power for the same reasons except that Japan relies on the sea lanes in the Indian Ocean for transporting over 90 percent of its oil. The Indian Navy, which consists of an aircraft carrier, fourteen submarines and fifteen major surface combatants, can project sea power not only in the Indian Ocean region but also in the South China Sea, the Mediterranean Sea, and the Persian Gulf. Its introduction of the Brahmos cruise missile and the possible lease of nuclear submarines from Russia will add significantly to the Indian navy's flexibility and striking power in South Asia.[7] Given the Indian navy's improved technology and military reach, Beijing views India as a strategic competitor, even though overall relations have improved considerably between the two nations since the end of the Cold War.

To counter Indian and U.S. naval influence in the Indian Ocean region China has quietly followed a strategy of military and economic cooperation with Pakistan, Myanmar, Bangladesh, Sri Lanka, and the Maldives.[8] China has been engaged in building a major naval base on Hangyi Island in Myanmar near the Irrawady river delta, which will soon be capable of hosting large Chinese naval vessels like nuclear attack submarines

(SSNs) and ballistic missile submarines (SSBNs).[9] Beijing has also been selling arms as well as conducting high-level talks about establishing a naval base with the government of Bangladesh, which would provide berthing facilities for the Chinese surface ships and submarines.[10] China is exploring the possibility of establishing an intelligence listening post in Sri Lanka as well as berthing facilities for its warships, and is engaged in negotiations with the Maldivian government to construct a submarine base at Marao Island.[11] Finally, in apparent contravention to its improved relations with the Indian government, China is building a deep-water port at Gwadar, Pakistan, that will not only provide China and Pakistan the ability to monitor the sea lanes from the Persian Gulf but also the ability to monitor U.S. naval activity in the Persian Gulf region and Indian naval activity in the Arabian Sea.[12] Berthing facilities can support a future Chinese Navy Carrier Battle Group (CVBG) as well as Chinese nuclear attack submarines, ballistic missile submarines, and guided missile submarines (SSGNs). The port supplies a maritime interface to the Sino-Pakistani Friendship 'Karakorum Highway' that opened China's access to Pakistan and southern Asia in the late 1960s.[13] Essentially, the port is a transshipment point for Chinese maritime trade. China can potentially transport crude oil to Western China via a future pipeline and thereby reduce its maritime vulnerability. Transit time and the costs of shipping crude oil from the Persian Gulf notwithstanding the $5.9 billion cost to build the oil pipeline, will be reduced.[14] Thus, the Gwadar Port offers a geo-economic and geo-strategic pivot to China and Pakistan.[15]

However, the realization of economic and strategic objectives of the Gwadar Port is largely dependent on upon the reduction of separatist violence in Baluchistan and Xinjiang. Baluch nationalists see the Gwadar Port project as an attempt at Chinese colonization and the capture of their natural resources.[16]

China also announced in December 2011 that it will establish its first overseas naval base in the Seychelles.[17] The base will be located on the island of Mahe, the country's largest island. According to Chinese authorities, the base will help China to conduct resource exploration of the Indian Sea and bolster its antipiracy efforts.[18]

China is alleged to have built a small intelligence gathering station on Great Coco Island in 1992. It is located at a crucial point in traffic routes between the Bay of Bengal and the Strait of Malacca. It is also an ideal location for monitoring Indian naval and missile launch facilities in the Andaman and Nicobar Islands to the south and movements of the Indian Navy and other navies throughout the eastern Indian Ocean. Overall, China's developing military and economic cooperation with the nations of South Asia will make it better positioned to protect its maritime energy transport routes during any future crisis.

In general, India views China's rise as threatening and encircling. Developments such as China's advances in the Indian Ocean, including an antipiracy role for the PLA navy in the Gulf of Aden; investments in militarily useful port facilities in Sri Lanka and Pakistan; and the continuing close security, economic, and other ties with India's principal adversary Pakistan are troubling according to analyst Jacques deLisle.[19] In fact, China's security relations with neighboring Myanmar and the construction of military and port facilities in the Indian Ocean is widely seen by Indian security strategists as an encirclement strategy designed to prevent or delay India's ability to challenge China's influence in South Asia.[20]

In summary, Beijing's developing relations and cooperation with the nations of South Asia will have strategic consequences for the India and Japan.

The PRINCE model predicts that China will build oil refineries in Pakistan based on a probability of 61 percent.[21]

Political, Economic, and Military Implications for India and Japan

Regionally, India seeks conventional security, energy security, prestige security, economic growth, and nuclear capability and nonproliferation while China seeks to expand multi-dimensional cooperative relations with all countries in the region.[22,23] In particular, India seeks to counter China's expansion into the region, to ensure the continued safe flow of goods and natural resources through the Bay of Bengal and beyond, particularly the area around the Malacca Strait and develop a nuclear triad.[24] Essentially, India is looking for, "mutual respectful partnerships that ensure the stability of the Indian Ocean."[25]

Japan's interest in South Asia and the Indian Ocean region is to ensure its energy security and to combat piracy.[26] Japan also seeks to build stronger trade relations with the countries in the region.[27] To ensure its energy security and to combat piracy, Japan constructed and opened the Djibouti naval base in July 2011. This is Japan's first, and sole, foreign military base since the Second World War. The base will provide Japan with a significant boost to its antipiracy efforts, while also helping to ensure the security of its shipping through the Indian Ocean's SLOCs.[28] Further, the new base will also play a major role in increasing Japan's strategic influence in the region.[29]

As explained above, China is building strategic relationships and developing a capability to establish a forward presence along the sea lines of communication in the Indian Ocean region that connect China to Africa and the Near East. Beijing continues to assert its sovereignty over Taiwan, countering Taiwan's efforts to expand its diplomatic ties regionally and with the United States, while working to guarantee South Asian neutrality in the event of a Sino-American military clash over Taiwan.[30]

China also seeks to minimize India's political alignment with the United States by encouraging U.S. engagement with Pakistan.[31] With Gwadar as a potential berthing location for Chinese warships, a Chinese naval base 240 miles from the Strait of Hormuz would provide not only strategic depth for China in the Indian Ocean region but may also directly threaten U.S. interests in the region. However, warming diplomatic ties between China and Pakistan will not necessarily occur at the expense of U.S.–Pakistani relations in the short-term because the United States regards Pakistan as an ally in its war against Taliban and al Qaeda.

Sino-Pakistani economic and trade ties, which are stronger than Sino-Indian economic and trade ties, are nurtured within the framework of military/strategic needs.[32] While China has been one of Pakistan's major suppliers of military aid and equipment, its energy activities in South Asia are unlikely to result in a Sino-Indian or Sino-Japanese economic competition for oil energy resources because the region does not produce crude oil for either India, or Japan. However, South Asia's economic significance is no less important to China, India or Japan because of its geographic location relative to the SLOCs. In this respect, South Asia is vital to China, India, and Japan's energy security. The PRINCE analysis indicates that China, India, and Japan will seek to minimize their economic risk along the SLOCs through continued military and economic cooperation with the nations of South Asia.

To establish regional security and stability, India has developed its diplomatic and economic relations with its regional neighbors. Further, India has been quietly modernizing its military including its navy. India has also increased its naval activity and presence in the Bay of Bengal. In particular, India established a strategic command in Andaman and Nicobar islands in August 2001 to safeguard its strategic interests in Southeast Asia and the Strait of Malacca.

As previously stated, Japan established a naval base at Djibouti in July 2011 to ensure the security of its shipping through the Indian Ocean's SLOCs.

India and Japan have also supported U.S. operations in the region. United States conducts military operations in support of political objectives such as counterterrorism and Weapons of Mass Destruction (WMD) non- and counter-proliferation. The United States also supports regional development and cooperation by providing military equipment, training and assistance, and information sharing, as well as promoting military outreach and exchange programs with the objective of building stronger relationships with regional military players such as Pakistan but not at the expense of India. India sees the United States as a potential ally in its ongoing struggle with Pakistan, while Pakistan sees the United States as a potential interloper.

To sustain international energy security, the United States Navy provides for the safe passage of national and international oil tankers transiting in the Indian Ocean region including the strategic chokepoints at the eastern and western approaches at the Strait of Malacca, the Strait of Bab-el-Mandeb (between the Red Sea and the Gulf of Aden) and the Strait of Hormuz. International navies from Asian and European countries are working in concert with the United States Navy to patrol at these strategic chokepoints. China has shown a willingness to take on a larger military role on the global stage not only to fulfill its international obligations to keep the SLOCs safe, but also to protect a more diverse set of national interests that are not defined by geography such as protecting economic interests, preventing terrorism, delivering humanitarian aid, and conducting other military operations other than war (MOOTW).[33] Beijing's deployment of a task force made up of two guided missile destroyers and a supply ship to participate in international antipiracy operations in the Gulf of Aden demonstrates its acceptance of its international responsibility to protect international commerce, not just its own.[34]

China seeks to expand military ties through joint military exercises and high-level exchanges at the defense minister level, regional level, military academy level and defense industrial agency level.[35] Beijing also seeks to achieve its political objectives of regional security and stability on substantive issues of mutual concern such as counternarcotics trafficking, smuggling, and terrorism and to expand cooperative relations with all nations in the region.[36]

In terms of energy security, China would seek to prevent hostile powers from interfering with the delivery (by land or sea) of its oil supply from the Near East and Africa.[37] Diversification strategies such as robust overland transportation systems between Yunnan and the Bay of Bengal and between Xinjiang Province in Western China and Gwadar, Pakistan as well as military strategies such as Chinese naval escorts from Near East ports to China would reduce China's energy insecurity in South Asia and substantially strengthen the Chinese military's ability to sustain intense military operations in the Indian Ocean region.[38] Further, if the Chinese Navy (People's Liberation Army Navy [PLAN]) established bases on the Indian Ocean littoral (such as Gwadar and Kyaukphyu, Myanmar) and link those forward bases to China by robust road and rail lines, America's ability to restrict oil shipments in the Indian Ocean region would be substantially diminished.[39] India would be inclined to bolster its naval capability by adding capital warships to its fleet such as aircraft carriers and nuclear submarines as well as increasing its naval presence in the Indian Ocean region. Unlike India, Japan would rely on the United States for maritime security and the protection of the SLOCs.

China–Pakistan–India–Japan Diplomatic/Political Relations

Sino-Pakistani diplomatic relations are robust even though Pakistan does not produce crude oil for China. Their relationship is enduring, multidimensional and deep-rooted underpinned by mutual trust.[40] Cooperation is strategic as evidenced by the fact that China has been a major arms supplier to Pakistan while Pakistan has supported China on domestic issues related to sovereignty and human rights. However, China's continued military and economic support for Pakistan which includes nuclear weapons and missile-related transfers is a major and ongoing source of friction between China and India.[41] Further, China's support for Pakistan's Kashmir position adds further discomfort to Sino-Indian bilateral relations.

Indo-Pakistani diplomatic relations have been strained by historical and political issues such as the partition of British India in 1947, the Kashmir dispute and numerous military conflicts. Attempts to improve bilateral relations through confidence-building measures (CBM) have been moderately successful since 2004 but are overshadowed by periodic terrorist attacks against India.

Japanese–Pakistani diplomatic relations have been strong since relations were first established in 1952. Pakistan also plays an essential role in fighting terrorism according to Japan's foreign ministry. Further, Japan sees Pakistan as a cornerstone in South Asia connecting politically, economically, and culturally to other countries in South Asia, Middle East, and Central Asia.

The construction of a seaport and naval base at Gwadar, oil refineries in Baluchistan and an oil pipeline along the Karakorum Highway linking China and Pakistan is unlikely to result in either Sino-Indian or Sino-Japanese diplomatic competition for oil energy resources because Pakistan does not produce crude oil for China, India, or Japan. However, India and Japan would be compelled to step up its diplomatic efforts to ensure that its strategic interests are promoted in South Asia because of Chinese competition for regional influence.

China–Pakistan–India–Japan Economic Relations

Sino-Pakistani economic relations are very good even though Pakistan does not produce crude oil for China. China is Pakistan's largest import trading partner with a 13.8 percent share in 2007.[42] Moreover, China is Pakistan's largest export trading partner after the United States, the United Arab Emirates, Afghanistan and the United Kingdom.[43]

Indo-Pakistani economic relations are weak as militarized tensions and territorial disputes between India and Pakistan have hamstrung economic and social development in both countries.[44]

Japanese–Pakistani economic relations are strong. Japan has been Pakistan's major source of economic assistance, a leading trading partner and an important source of foreign investment. In fact, Japan's economic assistance has played a very important role in the development of Pakistan's economic and social infrastructure.

The construction of a seaport and naval base at Gwadar, oil refineries in Baluchistan and an oil pipeline along the Karakorum Highway linking China and Pakistan is unlikely to result in either Sino-Indian or Sino-Japanese economic competition for oil energy resources because Pakistan does not produce crude oil for China, India, or Japan. However, Japan may be compelled to increase its economic aid and assistance to Pakistan to ensure that its economic interests are promoted because of rising Chinese influence.

China–Pakistan–India–Japan Military Relations

Sino-Pakistani military relations are robust even though Pakistan does not produce crude oil for China. Bilateral military relations have been characterized by high-level civilian and military visits and dialogues since 1956 and technical cooperation in conventional and strategic weapons programs. China's military relationship with Pakistan's military has often been regarded by the leaders of countries as all-weather and time-tested relationship.

Indo-Pakistani military relations are non-existent except for high-level defense talks in 2011 as diplomatic relations have been strained by historical and political issues.[45]

Japanese–Pakistani military relations have consisted of high-level ministerial and military consultations in 2010.[46] Rather, Pakistan enjoys an all-weather military relationship with Japan's archrival China.

The construction of a seaport and naval base at Gwadar, oil refineries in Baluchistan and an oil pipeline along the Karakorum Highway linking China and Pakistan is unlikely to result in either Sino-Indian or Sino-Japanese military competition for oil energy resources because Pakistan does not produce crude oil for China, India, or Japan. However, India would be compelled to bolster its naval capacity as well as increase its naval presence in the Indian Ocean region to ensure that its economic interests are protected because of increased Chinese naval presence in the region. In this respect, both China and India have overlapping strategic interests in South Asia and the Indian Ocean region which includes the protection of the sea lines of communication that commercial tankers transit to deliver crude oil to China and India. Japan has strategic interests in the Indian Ocean such as maritime security and the protection of the SLOCs. However, Japan relies heavily on the United States to accomplish these missions.

NOTES

1. Ministry of External Affairs, Government of India, "India–Kazakhstan Relations," July 2013, accessed October 8, 2013, http://www.mea.gov.in/Portal/ForeignRelation/India-Kazakhstan_Relations.pdf

2. INPEX Corporation, *Annual Report 2011: A Firm Commitment to Sustainable Energy*, 2011, 52.

3. Prince and Panda website, "PRINCE_SCORES.pdf," accessed October 17, 2013, http://www.princeandpanda.com.

4. Matthew Oresman, "Repaving the Silk Road: China's Emergence in Central Asia," in Joshua Eisenman, Eric Heginbotham, and Derek Mitchell, eds., *China and the Developing World: Beijing's Strategy for the Twenty-First Century* (Armonk, NY: M.E. Sharpe, 2007), 61–62; Bates Gill and Matthew Oresman, *China's New Journey to the West: China's Emergence in Central Asia and Implications for U.S. Interests* (Washington, DC: CSIS Press), August 2003, 13; and Thrassy N. Marketos, *China's Energy Geopolitics: The Shanghai Cooperation Organization and Central Asia* (New York: Routledge, 2009), 11.

5. Gill and Oresman, *China's New Journey to the West*, 1–2.

6. Ashley J. Tellis, "China and India in Asia," in Francine R. Frankel and Harry Harding, eds., *The India and China Relationship: What the United States Needs To Know* (Washington, DC: Woodrow Wilson Press, 2004), 158.

7. U.S. Department of State, "Background Notes: India," January 2009.

8. M. Matheswaran, "Air and Aerospace Partnership" in Swaran Singh, ed., *China–Pakistan Strategic Cooperation: Indian Perspectives* (New Delhi, India: Manohar Publishers, 2007), 198.

9. Vijay Sakhuja, "Naval Cooperation" in Swaran Singh, ed., *China–Pakistan Strategic Cooperation: Indian Perspectives* (New Delhi, India: Manohar Publishers, 2007), 217.

10. Sakhuja, "Naval Cooperation," 217–219.

11. Sakhuja, "Naval Cooperation," 220.

12. Sakhuja, "Naval Cooperation," 215–216.

13. W. Lawrence Prabhakar, "The Maritime Convergence" in Swaran Singh, ed., *China–Pakistan Strategic Cooperation: Indian Perspectives* (New Delhi, India: Manohar Publishers, 2007), 238.

14. Corey S. Johnston, "Transnational Pipelines and Naval Expansion: Examining China's Oil Insecurities in the Indian Ocean" (master's thesis: Naval Postgraduate School 2008), 83.

15. Prabhakar, "The Maritime Convergence," 238.

16. Tarique Niazi, "Gwadar: China's Naval Outpost on the Indian Ocean," *China Brief*, Vol. 5, Issue 4, February 28, 2005.

17. "Seychelles Chosen for China's First Overseas Naval Base," *Want China Times*, December 13, 2011, accessed October 8, 2013, http://www.wantchinatimes.com/news-subclass-cnt.aspx?id=20111213000041&cid=1101.

18. "Seychelles Chosen for China's First Overseas Naval Base," *Want China Times*, December 13, 2011, accessed October 8, 2013, http://www.wantchinatimes.com/news-subclass-cnt.aspx?id=20111213000041&cid=1101.

19. Jacques DeLisle, "Contested Terrain: China's Periphery and International Relations with Asia," in *E-Notes*, Foreign Policy Research Institute, March 2012, http://www.fpri.org/enotes/2012/201202.delisle.china-periphery.pdf.

20. Alan K. Kronstadt, *India–U.S. Relations*. CRS Report RL33529 (Washington, DC: Library of Congress, January 30, 2009), 19.

21. Prince and Panda website, "PRINCE_SCORES.pdf," accessed October 17, 2013, http://www.princeandpanda.com.

22. John W. Garver, "China's South Asian Interests and Policies," testimony, July 22, 2005, before U.S.–China Economic and Security Review Commission.

23. Xenia Dormandy, "India's Foreign Policy" (Paper presented at the India Forum, Madrid, Spain), November 5, 2007.

24. Dormandy, "India's Foreign Policy."

25. Karl Inderfurth and Bruce Riedel, "Breaking More Naan with Delhi: The Next Stage in U.S.–India Relations," *The National Interest* (Nov/Dec 2007): 56–62.

26. Bruno De Paiva, "Japan's National Involvement in the Indian Ocean Region," *Future Directions International*, July 29, 2011.

27. Paiva, "Japan's National Involvement in the Indian Ocean Region."

28. Paiva, "Japan's National Involvement in the Indian Ocean Region."

29. Paiva, "Japan's National Involvement in the Indian Ocean Region."

30. Garver, "China's South Asian Interests and Policies."

31. Garver, "China's South Asian Interests and Policies."

32. Madhu Bhalla, "Geopolitics of Economic Relations" in Swaran Singh, ed., *China–Pakistan Strategic Cooperation: Indian Perspectives* (New Delhi, India: Manohar Publishers, 2007), 251.

33. Alison A. Kaufman, *China's Participation in Anti-Piracy Operations off the Horn of Africa: Drivers and Implications* (Alexandria, VA: Center for Naval Analysis), July 2009, 1, 4, 7.

34. Kaufman, *China's Participation in Anti-Piracy Operations off the Horn of Africa*, 1.

35. Garver, "China's South Asian Interests and Policies."

36. Garver, "China's South Asian Interests and Policies."

37. Garver, "China's South Asian Interests and Policies."

38. Garver, "China's South Asian Interests and Policies."

39. Garver, "China's South Asian Interests and Policies."

40. U.S. Department of State, "Background Notes: Pakistan," March 2009.

41. Kronstadt, *India–U.S. Relations*, 19.

42. U.S. Department of State, "Background Notes: Pakistan," March 2009.

43. U.S. Department of State, "Background Notes: Pakistan," March 2009.

44. Kronstadt, *India–U.S. Relations*, 16.

45. Ministry of Defence, Government of India, *Annual Report 2011–2012*, 2012, 192.

46. Japanese Ministry of Defense, *Defense of Japan 2012*, 2012, 279.

5

The Middle East and North Africa (Near East)

In terms of the energy producing states for China, India, and Japan, the Middle East and North Africa region is comprised of Algeria, Iran, Iraq, Kuwait, Oman, Qatar, Saudi Arabia, the United Arab Emirates and Yemen. For the purpose of discussion and analysis, the key oil producing states for China, India, and Japan are Iran, Kuwait, Saudi Arabia, and the United Arab Emirates. Many observers see this region as the key area of Sino-Indian and Sino-Japanese competition because all have overlapping economic interests in the region—the acquisition of crude petroleum. Of particular interest are the countries of the Persian Gulf because they are principal suppliers of the world's oil and energy.[1] With respect to the Persian Gulf, Ashley Tellis opined that:

> The Persian Gulf will remain the main area where potential Sino-Indian rivalries over security and wealth will intersect, because cheap and abundant energy remains the motor for sustenance of high growth rates domestically-economic performance that both China and India define as essential.[2]

With respect to Japan's role in the Middle East, Geoffrey Kemp opined that:

> Japanese imports from the Middle East doubled between 2003 and 2007. Although those trade numbers are less significant than those for China or India, Japan will nevertheless be a competitor in the struggle for energy resources in the Middle East.[3]

In terms of energy trading, crude oil and petroleum imports from the Middle East and North Africa rank first in the world. China, India, and Japan trade heavily and are primarily dependent on the Near East for energy resources. In this respect, China, India, and Japan imports approximately 54 percent, 71 percent and 84 percent respectively from the Middle East and North Africa based on 2012 estimates from FACTS Global Energy. Thus, it is very likely that China, India, and Japan will be dependent and even increase their dependence on oil imports from the region by reason of necessity.

Further, although China's interest in the Near East is mainly economic, nevertheless, it has adopted a comprehensive, three-pronged strategy to promote its overall interests. First, China uses diplomacy to help Chinese businesses make inroads to the oil industry and markets of the Middle East and North Africa.[4] Second, China has "adopted an offend-no-one approach to regional politics in order to maintain an environment friendly to Chinese business interests."[5] Third, China exploits its soft power resources to cultivate goodwill and friendship with an eye toward protecting its long-term economic interests.[6] India and Japan also employ soft power strategies to cultivate relationships with the energy producing states in the region. However, India and Japan are not as adept as China in using soft power and do not have readily available instruments of soft power such as overseas infrastructure funding through significant national government financing and support. This is problematic for many Indian and Japanese oil companies trying to compete with Chinese oil companies.

The PRINCE model predicts that China will continue to explore for oil in, and import oil from, the Near East based on a probability of 94 percent or greater.[7]

NEAR EAST: POLITICAL IMPLICATIONS FOR INDIA AND JAPAN

In terms of diplomatic relations, China, India, and Japan generally have very good ties with the Near East especially with those countries that produce crude petroleum. Of the three countries, India has had the longest relationship with the region especially during World War I when India sent troops to fight or serve as noncombatants, or during World War II when India sent men and materials to support the allied war effort.[8] Shri Kamal Nath, India's minister of commerce and industry stated on March 25, 2006:

> India and the Members of the Gulf Cooperation Council have historic ties that go back not merely decades, but centuries.[9]

In comparison, China and Japan are newcomers. Kemp writes that Japan's political footprint in the greater Middle East is less apparent than China and India.[10] However, Japan will be a key player in the energy field.[11]

Although China is a relative newcomer to the Near East, its relations has rapidly expanded since 2000 to include a vast array of interactions.[12] Politically, China has created the Forum on China-Africa Cooperation (FOCAC) which includes North Africa and the China-Arab States Cooperation Forum (CASCF). The former will be discussed in detail in chapter 6.

CASCF is China's primary multilateral coordination mechanism with the Near East.[13] It was established in 2004 and represents the twenty-one members of the League of Arab States.[14] It meets every two years and all meetings are held at the ministerial level.[15] Through CASCF China actively negotiates with the Arab League for the inclusion of collective projects involving multiple Arab countries (e.g. railway projects, nuclear power projects, and the Dead Sea initiatives).[16]

China has also established a Middle East Issues Envoy in 2002 due to the urging by Arab States for China to become involved in the Middle East peace process and other concerns in the region.[17] Many Arab states perceive China to be a more balanced player in the Middle East peace process than other great powers.[18]

China has also established strategic partnerships to include all aspects of bilateral relations (political, economic, cultural, and military).[19] At the 2010 Tianjin ministerial meeting of CASCF, China established a strategic partnership with the League of Arab States.[20] To date, China has initiated strategic partnerships with the following countries in the Near East: Egypt (1999), Saudi Arabia (1999), Algeria (2004), and the United Arab Emirates (2012).[21]

Lastly, China's leadership has made numerous visits over the last two decades. The Jiang Zemin administration (Jiang Zemin, Zhu Rongji, and Li Peng) visited the following countries (number of visits in parentheses) between 1990 and 2002: Egypt (3), Algeria (2), Iran (1), Libya (1), Saudi Arabia (1) and Tunisia (1).[22] Between 2003 and 2012, the Hu Jintao administration (Hu Jintao and Wen Jiabao) visited: Egypt (3), Saudi Arabia (3), the United Arab Emirates (2) and Qatar (1).[23]

As to whether there will be political competition for energy resources, Tellis opined that:

> Sino-Indian competition over the Persian Gulf is likely to take to political rather than military form: within the ambit of a functioning energy market, both sides will seek supplemental arrangements for oil and gas; both sides will seek privileged relations with key regional states; both sides may even become modest suppliers of some military technologies, with China in

> particular becoming a potential supplier of disruptive technologies depending on its larger relations with the United States; and India will more likely than not work with the United States to ensure freedom of navigation along the Gulf transit routes.[24]

It is unknown whether Sino-Japanese political competition will rival Sino-Indian political competition for energy resources in the region but it is logical given Japan's heavy dependence on foreign oil especially from the Near East. In this respect, Japan imports 84 percent of its oil from the region. However, unlike China and India, Japan will not be a supplier of military technologies to the Gulf region or elsewhere where it seeks crude petroleum.

China–Iran–India–Japan Diplomatic/Political Relations

Indo-Iranian political relations have spanned for centuries and are generally very good. In the modern sense, India and Iran established diplomatic relations on March 15, 1950, three years after India obtained independence from the United Kingdom.

Since 1950 Indo-Iranian diplomatic relations have improved but were setback by a series of events such as: Iran's decision to join the Baghdad Pact in 1955, Iran's support of Pakistan in its 1965 war with India and Iran's Islamic Revolution of 1979.[25] The rise of Sunni Islamist forces in Afghanistan provided the unifying force to resume bilateral relations given Pakistan's support for the Sunni Islamists. This led to the deterioration of Pakistani–Iranian relations and put Pakistan in cooperation with one of Iran's regional enemies—Saudi Arabia.[26] Both supported the Afghan Northern Alliance in the fight with the Taliban and cooperated on the reconstruction of Afghanistan.

Indo-Iranian diplomatic relations picked up speed in 1997 when the reform-minded Mohammed Khatami became president of Iran.[27] India's motive for improving relations with Iran was driven by its need to diversify its energy supplies in light of the fact that the majority of its imported oil was shut off when Iraq invaded Kuwait in 1991. In this respect, Iraq and Kuwait supplied two-thirds of India's oil prior to Iraq's invasion of Kuwait.[28] It should be noted that the most important element in the rapprochement in Indo-Iranian diplomatic relations was predicated on overlapping economic interests in Central Asia.[29]

Japanese–Iranian political relations are generally good but have fluctuated since August 1929 when Japan opened a legation in Iran. Diplomatic relations were suspended in April 1942 and reopened in November 1953 after the Second World War.[30] While bilateral relations date back to many centuries, Japan's relations with Iran have been somewhat rocky because of U.S. pressure and Japan's own concerns about Iran's behavior regarding the pursuit of nuclear technology.[31] The United States and Japan have been allies since the end of the Second World War and have bilateral security agreements which complicate Japan's political relations with Iran. Conversely, complicating U.S.–Japanese relations is Japan's heavy dependence on Iranian oil in light of the fact that Japan relies entirely on imported oil for economic sustenance. Japan imports approximately 12 percent of its crude petroleum from Iran making Iran one of Japan's major oil suppliers.[32] In fact, Iran is Japan's fourth-largest oil supplier after the United Arab Emirates, Saudi Arabia, and Kuwait. Thus, it behooves Japan to maintain good relations with both Iran and the United States but doing so can be very tricky.

Sino-Iranian political relations date back to pre-Islamic times. Their shared historical experiences of empire and autonomy as well as Western colonial power domination further unite them and form a significant part of their relationship.[33]

One of the main pillars of the bilateral relationship is oil and gas. China imports approximately 10–11 percent of its crude petroleum from Iran making Iran one of its major oil suppliers in the Near East according to FACTS Global Energy. Because of its reliance on Iranian oil and gas, China has invested into Iran's infrastructure such as dam and ship construction, and seaport and airport development.[34] China is also committed to modernizing Iran's oil and gas sector in order to secure energy resources.

China's energy trading with Iran is unlikely to result in either Sino-Indian or Sino-Japanese political competition for energy resources such as crude petroleum unless China pursued or acquired exclusive rights to Iran's oil resources to the detriment of either India or Japan. Further, if either India or Japan meddled in the internal affairs of Iran such as nuclear technology issues, it is very likely that Iran would cease exporting oil to either country. However, the likelihood of either India or Japan interfering in Iran's internal politics is nonexistent because they generally don't mettle internally.

China–Kuwait–India–Japan Diplomatic/Political Relations

Indo-Kuwaiti political relations are warm and friendly and span for centuries. In fact, geographical proximity, historical trade links, cultural affinities and the presence of a large number of Indian expatriates continue to sustain and nurture this close relationship.[35]

In terms of energy trading, India imports approximately 10–11 percent of its annual crude petroleum from Kuwait making Kuwait a major oil supplier to India.[36]

Japanese–Kuwaiti political relations are very good. Diplomatic relations were established in December 1961 and have been marked by numerous bilateral official visits in recent years.[37] In terms of energy trading, Japan imports approximately 8 percent of its annual crude petroleum from Kuwait making Kuwait a major oil supplier to Japan.

China and Kuwait established diplomatic relations in March 1971 making Kuwait the Gulf Cooperation Council country with the longest diplomatic relationship.[38] Their relationship has been characterized by bilateral, high-level official visits and bilateral cooperation in trade, energy, finance, and infrastructure development.[39]

In terms of energy trading, China imports approximately 3.5 percent of crude petroleum annually from Kuwait according to FACTS Global Energy. Thus, Kuwait is not a major supplier of oil to China.

China's energy trading with Kuwait is unlikely to result in either Sino-Indian or Sino-Japanese political competition for energy resources such as crude petroleum unless China pursued or acquired exclusive rights to Kuwait's oil resources to the detriment of either India or Japan. It should be noted that unlike India and Japan, China is not heavily dependent on Kuwait's energy resources because Kuwait is a minor oil supplier to China.

China–Saudi Arabia–India–Japan Diplomatic/Political Relations

Indo-Saudi diplomatic relations were established in 1947 and are primarily based on trade.[40] Relations were generally good throughout the Cold War but India's close relations toward the Soviet block and Saudi Arabia's relations to Pakistan prevented the relationship from further developing.[41] Relations improved in the early 1980s but were thwarted by Cold War tensions arising from their opposing views on the Soviet invasion of Afghanistan. India supported the Northern Alliance as a means of combating Pakistani influence in Afghanistan while Saudi Arabia supported the mujahedeen and subsequently the Taliban.[42]

In the 1990s relations began to improve again in the wake of the Cold War and first Gulf War.[43] India began to focus on economic development. Bilateral visits and exchanges occurred since the 1990s leading up to the signing of the Delhi Declaration of 2006, which included a mutual agreement to deepen and broaden economic ties while cooperating in combating international crime and ensuring the stability of free trade.[44] Further, it appeared that Saudi Arabia was tilting diplomatically toward India to the detriment of Pakistan after King Abdullah of Saudi Arabia suggested that India should have observer status in the Organization of the Islamic Conference.[45]

High-level reciprocal diplomatic contacts continued in recent years and have resulted in the signing of many agreements between India and Saudi Arabia such as an agreement to cooperate in the fields of hospital consultancy and management, the relaxation of visa restrictions for business travelers, and funding for infrastructure investment.[46] India and Saudi Arabia also signed a number of memorandums of understanding (MOUs) on foreign office consultations, scientific and technical cooperation, cultural cooperation, commerce, sports, education, the peaceful use of outer space, the protection of investments, media cooperation, and extradition.[47] Overall, diplomatic relations are cordial and friendly today.[48]

In terms of energy trading Saudi Arabia is India's largest supplier of crude petroleum. India imports almost 20 percent of its oil from Saudi Arabia.[49] In fact, the importation of crude oil from Saudi Arabia to India forms a major component of their bilateral trade.[50]

Japanese–Saudi diplomatic relations were established in June 1955 and are generally very good. Reciprocal visits by members of the Saudi royal and Japanese imperial families in addition to premier and cabinet-level visits have strengthened diplomatic relations between the countries.[51]

High-level reciprocal diplomatic contacts between Japan and Saudi Arabia have resulted in numerous bilateral treaties and agreements on economic and technical cooperation in 1975, the Comprehensive Partnership toward the Twenty-First Century and the Japan–Saudi Arabia Cooperation Agenda in 1998.[52] The latter initiative was significant because it focused on human resources development, the environment, health sciences and technology, culture and sports, and investment and joint ventures.[53]

In terms of energy trading Saudi Arabia is Japan's largest supplier of crude petroleum. Japan imports approximately 31 percent of its oil from Saudi Arabia.[54] Thus, Saudi Arabia is Japan's essential trading partner as well as its largest trading partner in the Middle East.[55]

Sino-Saudi diplomatic relations were established in 1990.[56] While diplomatic relations have developed since 1990, economics such as oil and gas trading and investment have formed the crux of the Sino-Saudi relationship.[57] In fact, Saudi Arabia is China's largest source of crude oil accounting for approximately 20 percent of China's annual crude oil imports in 2012 according to FACTS Global Energy. Moreover, China is Saudi Arabia's most important "market for its 'distressed' medium-grade oil—a viscous, acidic, often sulfurous product that Saudi Arabia has in abundance but that has few takers on the international market."[58] Consequently, "Saudi Arabia has been working to develop China's capacity to develop and use Saudi heavy crude, investing in two refineries along China's coast": the upgrading of the Quanzhou refinery in Fujian Province and the development of the Qingdao refinery in Shandong province.[59] Further, in 2006, China and Saudi Arabia agreed to jointly construct an oil storage facility on China's Hainan Island and Saudi Arabia invited Chinese firms to participate in infrastructure development worth $624 billion.[60]

In terms of developing political ties, Chinese President Hu Jintao and Saudi Arabian King Abdallah conducted high-level exchange visits to each other's countries in early 2006, in which President Hu laid a four-point proposal for enhancing Sino-Saudi relations. One point called for the mutual safeguard of each other's national sovereignty and territorial integrity and the continued support and cooperation on international and regional affairs.[61] Since 2006, the two countries frequently made bilateral high-level visits and deepened mutual political trust.[62]

China's energy trading with Saudi Arabia is very unlikely to result in either Sino-Indian or Sino-Japanese political competition for energy resources such as crude petroleum unless India or Japan meddled in the internal affairs of Saudi Arabia. It is not likely that either India or Japan will do so not only because of their robust trade relationship with Saudi Arabia but also their oil dependence. In this respect, Saudi Arabia is India and Japan's chief crude petroleum supplier.

China–United Arab Emirates–India–Japan Diplomatic/Political Relations

Indo-Emirati diplomatic relations have existed for millennia based on deep cultural, religious and economic ties long before the UAE was formed in 1971.[63] Bilateral ties have carried over to the current relationship, which was grounded in the oil boom of the 1970's when the Persian Gulf region's need for labor led to many Indians traveling to the UAE.[64] In fact, the UAE is home to an Indian expatriate community of more than 1.75 million—the largest expatriate community in the UAE.[65] The Indian community has played a major role in the economic development of the UAE.[66]

High-level reciprocal diplomatic contacts between India and the UAE have resulted in numerous bilateral treaties, agreements and MOUs on cultural cooperation (1975), civil aviation (1989), double taxation avoidance (1992), combat trafficking in narcotic drugs and psychotropic substances (1994), extradition (1999), mutual legal assistance in criminal and civil matters (1999), juridical and judicial cooperation in civil and commercial matters (1999), information cooperation (April 2000), channel carriage (2000), defense cooperation (2003), manpower sourcing (2006), political consultations (June 2011), transfer of sentenced persons (November 2011), security cooperation (November 2011), customs cooperation (April 1, 2012), streaming the admission of Indian contract workers by way of an electronic contract registration and validation (April 2012), and consular affairs.[67]

In terms of energy trading the UAE is one of India's major suppliers of crude petroleum. India imports approximately 9 percent of its oil from the UAE according to FACTS Global Energy.

Japanese–Emirati diplomatic relations were established in May 1972.[68] Like other countries in the region, ties are grounded in trade. In this respect, the UAE is Japan's second-largest trading partner in the Middle East after Saudi Arabia.[69] In fact, the UAE is Japan's second-largest oil supplier after Saudi Arabia.[70] In 2012, the UAE supplied approximately 22 percent of Japan's imported oil according to FACTS Global Energy.

In addition to robust trade ties, Japan and the UAE have reciprocated politically through high-level diplomatic exchanges and visits by Japan's past and present prime ministers and minister of foreign affairs and the UAE's Crown Prince Abu Dhabi H.H. Sheikh Mohammed bin Zayed Al Nahyan, the minister of economy and minister of foreign affairs.[71]

Sino-Emirati diplomatic relations were established in November 1984.[72] Unlike other countries in the region, the UAE is only a minor supplier to China accounting for 3.2 percent of China's oil imports in 2012 according to FACTS Global Energy.

While bilateral cooperation has occurred on many levels such as culture, education, trade, the economy, energy, and finance, China and the UAE have also engaged each other through high-level diplomatic exchanges in recent years. In addition, diplomatic exchanges have occurred at the local government level between the Chinese provincial leaders and the UAE's cabinet ministers. Such exchanges have improved diplomatic ties and mutual political trust between China and the UAE.[73]

China's energy trading with the UAE is very unlikely to result in either Sino-Indian or Sino-Japanese political competition for energy resources such as crude petroleum because the UAE is a very minor oil supplier to China. Essentially, the UAE is inconsequential to China's energy security given that the UAE accounts for approximately 3.2 percent of China's oil imports according to FACTS Global Energy. In addition, India and Japan are unlikely to damage their political relationship with the UAE such as meddling in the UAE's domestic politics given that UAE is a major oil supplier to them.

NEAR EAST: ECONOMIC IMPLICATIONS FOR INDIA AND JAPAN

China's interest in the Middle East and North Africa is primarily economic: energy trading, energy cross-investment activities and the international trade of goods and services. In terms of energy trading, the Middle East and North Africa is China's major crude oil supplier. Moreover, in terms of cross-investment activities, the Middle East and North Africa have provided China with upstream, downstream and infrastructure development opportunities in Algeria, Egypt, Iran, Iraq, Libya, Oman, Qatar, Syria, Tunisia, and Yemen. See appendix B. Further, China has sought capital from the Middle East and North Africa to invest in the development of China's oil industry. Finally, in terms of goods and services, the Middle East and North Africa is a major import market for lower- and intermediate-end consumer products, including machinery and electronic products, clothes, textiles, shoes, tires, and handbags.[74]

China's economic interests (resources and markets) in the region has dramatically increased from 1990 to 2012.[75] China has actively encouraged companies in energy, construction, and telecommunications to pursue business opportunities in the region.[76] To facilitate trade, China launched two special economic zones (SEZs) in Egypt (1994) and Algeria (2000).[77] The Egyptian Suez Canal SEZ is ideal because it is strategically located near port facilities on the Suez Canal.[78]

China and the Gulf Cooperation Council (GCC) also signed a framework agreement on economic, trade, investment, and technological cooperation and announced the launch of free trade agreement negotiations in July 2004.[79] The proposed agreement would include goods, services, and investment.[80]

Although China depends on the Middle East and North Africa for almost 50 percent of its crude oil, it is unlikely that China's crude oil energy trading with and cross-investment activities in the Middle East and North Africa will shift the balance of economic power in China's favor to the detriment of India and Japan in the near future not only because of India and Japan's strong economic ties to the region but also because of China's inability to project naval power beyond the Pacific Ocean to protect the vital sea lines of communication of which the Near East is so heavily dependent on for its international trade. In this respect, the United States is the only power in the world that can ensure the safe and timely flow of crude oil by ocean tankers originating from the Middle East and North Africa, of which China, India, and Japan and the Near East are economically dependent.

As the fastest growing economy in the world, it is possible that China's high demand for crude oil from the Middle East and North Africa could contribute to a rise in international oil prices, which may affect the economies of India and Japan. In this respect, the Organization of Petroleum Exporting Countries (OPEC) could use China's high demand for crude oil as an excuse to cut oil production and raise world oil prices, which, in turn, may have dire economic consequences for high oil importers like India and Japan. Obviously, petroleum prices would be higher in India and Japan as a result of strategic competition for oil energy resources in the Near East. Japan would be especially affected given that it is almost entirely resource dependent on imported oil from the Near East.

India's primary interest in the Near East is economic: to secure long-term access to oil and natural gas resources in the Persian Gulf and the Indian Ocean littoral in light of a growing economy.[81] India is also interested in investment opportunities, consumer good sales, tourism, and education.[82] In terms of energy investments, India has acquired oil exploration and/or production stakes in Egypt, Iran, Iraq, Libya, Oman, Syria, and Yemen. See appendix C. In terms of crude oil trading, India imports oil from Algeria, Egypt, Iran, Iraq, Kuwait, Libya, Qatar, Saudi Arabia, Syria, the United Arab Emirates, and Yemen with the major oil suppliers being Iran, Iraq, Kuwait, Saudi Arabia, and the United Arab Emirates according to FACTS Global Energy.

India also accounts for the largest number of expatriates in the Gulf region.[83] Approximately 4.5 million Indians work in the GCC countries—mostly in the UAE, Bahrain, Saudi Arabia, and Qatar—making them essential for the economic growth and well-being of the Gulf region.[84] The mutual dependence of India and the Gulf region, given their interdependence of labor and energy, has drawn them closer together.[85]

Japan's interest in the Near East is also economic.[86] Given Japan's great dependence on energy resources like oil and gas, its economic focus on the region is appropriate and adequately sums up how its role has evolved. However, Japan has pursued a very cautious policy on the most sensitive political issues such as Iran's pursuit of nuclear technology because it is fearful of incurring U.S. wrath.[87]

In terms of energy trading, Japan imports approximately 84 percent of oil from the Near East according to FACTS Global Energy. Japan imports crude petroleum from Iran, Iraq, Kuwait, Oman, Qatar, Saudi Arabia, the United Arab Emirates, and Yemen. Of these, Iran, Kuwait, Qatar, Saudi Arabia, and the United Arab Emirates are the chief suppliers.

In terms of energy investments, Japan has energy stakes in Algeria, Egypt, Iraq, Libya, Oman, Qatar, and the United Arab Emirates. See appendix D. Part of Japan's strategy for securing oil supplies from the region is to encourage investment in oil supplying countries especially in the Gulf region.

In terms of exports to the Near East, Japan mainly trades automobiles and consumer electronics.[88] Japan also trades general machinery, iron and steel, and textiles with Saudi Arabia and the United Arab Emirates as its top trading partners in the region.

China–Iran–India–Japan Economic Relations

Indo-Iranian economic relations can be traced back to ancient times but formally to 1950 when independent India and Iran first established diplomatic relations.[89] India's commercial and trade ties are largely buoyed by India's import of Iranian crude oil.[90] In fact, Iran is India's fourth-largest supplier of crude oil, so it is not surprising that energy plays an important part of their relationship.[91]

In terms of overall trade, Indo-Iranian trade was $13.4 billion in 2010–2011. Indian exports were $2.5 billion and imports were $10.9 billion.[92] While Iran exports crude petroleum to India, India exports to Iran rice, machinery and instruments, metals, primary and semi-finish iron and steel, drugs/pharmaceuticals, fine chemicals, processed minerals, manmade yarn and fabrics, organic/inorganic/agro chemicals, and rubber manufactured products.[93] In addition, India and Iran have engaged each other regarding commercial development such as the Chabahar Port and railway project.[94]

Although Japanese–Iranian economic relations have been in existence since formal diplomatic relations were established in April 1942, bilateral trade informally can be traced back many centuries.[95] Japan's trade ties with Iran are largely characterized by energy imports of which Iran supplies approximately 10–11 percent of Japan's total oil imports annually.[96] Thus, energy is central to Iranian–Japanese trade ties.

In terms of overall trade, Japanese–Iranian trade was $14.5 billion in 2011.[97] Japanese exports were $1.7 billion and imports were $12.8 billion.[98] While Iran exports crude petroleum and natural gas to Japan, Japan exports general machinery, motor vehicles, and iron and steel to Iran.[99]

In terms of energy investments in Iran, the Japanese oil company Japanese Petroleum Exploration Co. Ltd. (JAPEX) has a 20 percent interest to develop the Soroosh and Nowrooz oilfields offshore in the northern Persian Gulf.[100] Moreover, Japan's Inpex Corporation previously had a 75 percent interest to develop the Azadegan oil field in February 2004 but its holding was reduced to 10 percent in November 2006 when the Azadegan agreement was dissolved.[101] The dissolution jeopardized Japanese plans to ensure energy security by increasing the stakes of Japanese-owned companies in foreign energy assets.[102]

Sino-Iranian economic relations are robust as evidenced by bilateral trade of $30–40 billion in 2012.[103] Oil and gas trading establish the foundation for the economic partnership given Iran's vast energy resources and China's expanding energy needs.[104] In fact, China is Iran's largest oil customer and largest trading partner.[105]

In terms of bilateral trade, China exports power generation, mining, transportation equipment, and consumer goods such as electronics, auto parts, and toys to Iran while Iran exports crude oil and natural gas to

China. Iran is also the largest market for Chinese arms exports.[106] China has been supplying Iran with weapons since the mid-1980s.[107] Such weapons have included missiles and missile technology, including anti-ship missiles and surface-to-surface cruise missiles including several hundred Silkworm missiles.[108]

China is also a major infrastructure project developer in Iran. Over one hundred Chinese companies operate in Iran, primarily in dam and ship construction, energy, and seaport and airport development.[109] Projects include small business ventures in fish canneries, sugar refineries, paper mills, construction of Tehran's subway system, and the development of a $3 billion oil refinery.[110]

Although China has robust hydrocarbon energy interests in Iran such as major crude oil imports from Iran and ongoing upstream sector activities in Iranian oil fields, it is unlikely that Chinese energy activities will have either short-term or long-term economic consequences for either India or Japan unless Iran uses China's high demand for crude oil as an excuse to cut oil production and raise oil prices. However, Iran has the potential to disrupt the flow of commercial oil transiting through the Persian Gulf, which would have dire economic consequences for India and Japan in light of the fact that both countries import almost over 70 percent of their crude oil annually from the region. Iran could attack commercial oil tankers in the Persian Gulf and in the Strait of Hormuz as it has done so in the past during the Tanker War (1984–1988) and the United States Navy would be called out to respond militarily. Further, this scenario could escalate into a military conflict which would affect not only the economy of the Gulf States but also the economies of China, India, and Japan given their high dependence on Persian Gulf oil.

China–Kuwait–India–Japan Economic Relations

Indo-Kuwaiti economic relations are very good. Historically, bilateral relations have always had an important trade dimension.[111] India has consistently been amongst the top ten trading partners of Kuwait while Kuwait has been an important partner in India's quest for energy security, annually providing 10–11 percent of India's crude oil.[112]

In terms of bilateral trade, 2011 was an important year for both countries as trade reached $12.27 billion.[113] Major items imported by India from Kuwait other than crude petroleum and natural gas included mineral fuels, mineral oils and products of their distillation; bituminous substances; mineral waxes, organic chemicals, plastic and articles thereof, iron and steel, and aluminum.[114] Major items exported to Kuwait from India included articles of iron or steel, nuclear reactors, boilers, machinery and mechanical appliances and parts thereof; cereals, meat and edible meat offal, electrical machinery and equipment and parts thereof; sound recorders and reproducers, television image and sound recorders, and reproducers and parts.[115]

India and Kuwait also have bilateral business ventures. India has housing and financing, insurance, bridge and roofing, aviation and pharmaceutical companies in Kuwait while Kuwait has provided direct investment and assistance in India's petrochemical industry, including chemical production and downstream refining.[116] Further, Indian nationals make up the largest expatriate community in Kuwait representing practically all segments of Kuwaiti society.[117] Their contribution to the development and advancement of Kuwait is noteworthy according to the Indian Ministry of External Affairs.[118]

Japanese–Kuwaiti economic relations are generally very good. In 2011, bilateral trade was approximately $14.5 billion.[119] In terms of bilateral trade, Japan exports to Kuwait automobiles, machinery and electrical equipment while Kuwait exports to Japan crude oil and petroleum projects.[120]

While their trade relationship is less significant compared to Japan's relations with other Gulf States, Japan is Kuwait's largest purchaser of oil.[121] In this respect, Kuwait provides roughly 7–8 percent of Japan's annual oil imports and oil exports to Japan account for approximately 20 percent of Kuwait's total annual crude exports.[122]

Sino-Kuwaiti economic relations predate their diplomatic relations. Trading began in 1955 (sixteen years before diplomatic relations were established) and appear robust. In fact, in 2008, bilateral trade was $6.78 billion.[123]

In terms of bilateral trade, China exports to Kuwait textiles, electrical appliances, toys, gifts, foodstuffs, and cattle while Kuwait exports to China crude oil, urea, and scraps.[124]

In terms of energy investments, Kuwait Petroleum International (KPI), a subsidiary of KPC, signed a memorandum of understanding with Sinopec Corp. to jointly construct an $8–$9 billion oil refinery and a petrochemical complex in China.[125] The planned facility would include a refinery that could process fifteen million tons of Kuwaiti crude oil a year (three hundred thousand barrels per day) and an ethylene cracker unit with an annual production capacity of one million tons.[126] The project was expected to be operational in 2013.

China's energy trading with Kuwait is very unlikely to result in either Sino-Indian or Sino-Japanese economic competition for energy resources such as crude petroleum because Kuwait is a very minor oil exporter to China. In this respect, Kuwait supplies China with roughly 3.5 percent of its annual oil imports. Kuwait is inconsequential to China in terms of China's energy security.

China–Saudi Arabia–India–Japan Economic Relations

Indo-Saudi economic relations are very good. In fact, Indi-Saudi ties are based primarily on trade.[127] Saudi Arabia is India's fourth-largest trading partner and in 2010–2011, bilateral trade was $25.6 billion.[128]

In terms of bilateral trade, India exports to Saudi Arabia basmati rice, textiles and garments, and machinery while Saudi Arabia exports to India organic and inorganic chemicals, metal scrap, leather, gold, and oil.[129]

Energy trading is a large component of Indo-Saudi bilateral trade. Saudi Arabia is India's largest supplier of crude oil accounting for almost 20 percent of its needs.[130]

In recent years, India and Saudi Arabia have sought to broaden their trade relationship through increased investment and partnerships.[131] Many Indian companies have established joint venture projects or wholly-owned subsidiaries in Saudi Arabia.[132]

According to the Saudi Arabian General Investment Authority (SAGIA), more than 190 Indian companies have obtained licenses to set up fully-owned or joint venture projects with Saudi nationals.[133] Conversely, Saudi Arabian companies have been investing in India amounting to $33.1 million between April 2000 and August 2011.[134] Further, many Indo-Saudi joint ventures or Saudi-owned companies operate in India, in diverse fields such as paper manufacturing, chemicals, computer software, granite processing, industrial products and machinery, cement, and metallurgical industries.[135]

Japanese–Saudi economic relations are also very good. In fact, Saudi Arabia is considered an essential trading partner to Japan and is Japan's largest trading partner in the Near East.[136] In 2008, bilateral trade was $60 billion.[137]

In terms of bilateral trade, Japan exports to Saudi Arabia automobiles, machinery and equipment, and metals while Saudi Arabia exports to Japan crude oil, petroleum products, and liquefied petroleum gas.[138]

Petroleum energy is also a large component of Japanese–Saudi bilateral trade. Saudi Arabia is Japan's largest supplier of crude oil accounting for almost 28 percent of Japan's overall annual oil imports.[139]

Although Japan does not provide financial assistance because Saudi Arabia's per capita GNP is high, Japan provides technical cooperation on a project basis; accepts trainees in sectors such as communications, broadcasting, mining, and manufacturing; sends experts to the country; and offers other forms of assistance.[140] Both countries also participate in a trainee-expert exchange to promote technical advancement.[141] As of 2004, 1,750 trainees were sent to Japan and 75 Japanese experts were sent to Saudi Arabia.[142]

Sino-Saudi economic relations are robust. China initiated a strategic partnership with Saudi Arabia in 1999.[143] In fact, Saudi Arabia is China's largest trading partner in West Asia and Africa.[144] In 2012, bilateral trade was $74 billion.[145]

In terms of bilateral trade, China exports to Saudi Arabia cheap consumer goods including electronics, textiles, and food while Saudi Arabia exports to China crude oil and non-oil products such as chemicals used to make plastic products.[146]

The major component of their economic relationship is energy trading. In fact, economic ties centering on energy resources is the primary driver of their relationship.[147] In this respect, Saudi Arabia is China's largest supplier of crude oil in the world, followed by Angola and Iran.[148] Saudi Arabia's oil exports to China account for over 20 percent of China's overall annual oil imports according to FACTS Global Energy.

Bilateral trade has grown significantly since diplomatic relations were established in 1990. Because of the increased bilateral trade, Saudi Arabia has become a major investor in China.[149] Saudi Arabia has invested $750 million of the $3 billion needed to construct a petrochemical complex in southeastern Fujian Province that will process 8 million tons of Saudi crude oil.[150] Saudi Arabia, along with several members of the Organization of the Petroleum Exporting Countries (OPEC) also agreed to finance and build a new oil refinery in Guangzhou at a cost of $8 billion.[151]

Chinese investment in Saudi Arabia is paltry in comparison to Saudi investment in China. Activities of the Chinese National Oil Companies (NOCs) in Saudi Arabia are very much limited to engineering services, such as pipeline and well repair, seismic data collection, and natural gas projects.[152] Although Saudi Arabia does not allow foreign companies (including Chinese companies) to invest in its upstream (exploration and production) oil sector, it has allowed them to invest in the upstream gas sector and refinery sector.[153] Despite these restrictions, bilateral trade is likely to grow between China and Saudi Arabia.

China's energy trading with Saudi Arabia is unlikely to result in either Sino-Indian or Sino-Japanese economic competition for energy resources such as crude petroleum unless China and Saudi Arabia pursue an exclusive strategic partnership that disfavors or excludes India or Japan. An exclusive relationship that disfavors or excludes India or Japan is unlikely because Saudi Arabia has diversified its refining capacity globally according to the U.S. Energy Information Administration (EIA).[154] Saudi Arabia has approximately 2 million barrels per day of refining capacity overseas through joint and equity ventures in facilities in the United States, China, South Korea, Japan, and the Philippines.[155]

Despite Saudi Arabia's oil refining diversification, U.S. engagement of Saudi Arabia and the Near East, which includes a naval presence in the Arabian Gulf and the Indian Ocean, provides for regional stability and the maintaining balance of power. The United States, more than any world power, has the resources and capabilities to prevent any aspiring world power from influencing the region in such a manner that would economically disfavor or exclude regional powers like India or Japan especially in energy trading. Therefore, it is critical that the United States remains engaged in the region.

China–United Arab Emirates–India–Japan Economic Relations

Indo-Emirati relations are primarily driven by economics.[156] In fact, India is the UAE's largest trading partner.[157] In 2010–2011, bilateral trade was $67 billion.[158]

In terms of bilateral trade, India exports to the UAE petroleum products; precious metals; stones, gems and jewelry; minerals; food items (cereals, sugar, fruits and vegetables, tea, meat, and seafood); textiles (garments, apparel, synthetic fiber, cotton, yarn); and engineering and machinery products and chemicals while the UAE exports to India petroleum and petroleum products; precious metals; stones, gems and jewelry; minerals; chemicals; and wood and wood products.[159]

Crude petroleum is a large component of bilateral trading. The UAE was the fifth-largest exporter to India accounting for 8 percent of India's annual petroleum imports in 2007.[160] India annually imports approximately 9 percent of its oil from the UAE according to FACTS Global Energy making the UAE a major oil exporter to India.

Beyond oil trading, the UAE is one of India's largest investors. Total FDI to India from the UAE is estimated to be $2.2 billion.[161] UAE's investments in India are concentrated in five sectors: energy, computer programming, services, construction, and tourism and hotels.[162] Conversely, India's supplies the UAE with foreign labor. Currently, 33 percent of the UAE's population and 50 percent of its workforce are Indian, about 1.4 million people.[163]

Japanese–Emirati economic relations are robust. Bilateral trade was $57.2 billion in 2008.[164] The UAE is also Japan's second-largest trading partner in the Middle East after Saudi Arabia.[165] Further, the UAE is Japan's second-largest oil supplier after Saudi Arabia.[166]

In terms of bilateral trade, Japan exports to the UAE automobiles, electric items, machinery, and textiles while UAE exports to Japan crude oil and natural gas.[167] Japan has also been involved in the UAE's private sector such as establishing a distribution center for Mitsubishi Fuso Truck and Bus Corporation in Dubai in 2007 and constructing massive resort islands shaped like palm trees and a map of the world.[168]

Sino-Emirati economic relations are robust. Bilateral trade was $36 billion in 2011.[169] China is also considered a major trading partner of the

UAE. China's presence is visible in the form of the Dragon Mart, a huge shopping and trading hub covering eighty-two square miles in Dubai.[170] Further, China initiated a strategic partnership with the UAE in 2012.[171]

In terms of bilateral trade, China exports to the UAE textiles, clothes, industrial and metallic products, handicrafts, and machinery while the UAE exports to China aluminum, chemical fertilizer, petroleum, and polyethylene.[172] However, the UAE is only a very minor oil supplier to China. The UAE annually exports to China less than 3.2 percent of China's oil according to FACTS Global Energy.

China's energy trading with the UAE is very unlikely to result in either Sino-Indian or Sino-Japanese economic competition for energy resources such as crude petroleum because the UAE is a very minor oil supplier to China. Essentially, the UAE is more or less inconsequential to China's energy security compared to Saudi Arabia's or Iran's.

NEAR EAST: MILITARY IMPLICATIONS FOR INDIA AND JAPAN

China, India, and Japan have military relations with the Near East but in various degrees. Military relations can be bilateral or multilateral security talks, dialogues and consultations, military attaché representation, civilian official to civilian official visits and exchanges, military-to-military visits and exchanges, military exercise participation, peacekeeping operations, weapons sales, defense cooperation and formal security agreements. China has no known security agreements but has conducted anti-piracy patrols in the Gulf of Aden, sold weapons to Iran and Saudi Arabia, and stationed defense attachés in key oil producing countries. Between 2001 and 2010, China performed high-level military exchanges with every country in the region except Iraq.[173] China's highest volume of exchanges (number in parentheses) were with Egypt (29), Syria (10), and Tunisia (8).[174] India has cooperated with key oil producing states on counterterrorism, conducted antipiracy patrols in the Gulf of Aden, conducted joint military exercises with Iran, and posted military attachés in the Gulf Region. India also has participated in UN peacekeeping operations in Lebanon (UNFIL) and Syria (UNIDOF). Japan has also assigned defense attachés to the region, conducted joint military exercises, conducted anti-piracy patrols in the Gulf of Aden, participated in UN peacekeeping operations such as in the Golan Heights and Iraq, and signed a memorandum of defense exchange with Bahrain in April 2012.

China's oil energy trading and activities in the Middle East and North Africa is unlikely to lead to a military competition for strategic energy resources between China, India, and Japan because of the strong military relationships that exist between the United States and the countries in the Near East except Iran, Libya, and Syria. While the United States is not an official ally of India, the Congressional Research Service stated that:

> The United States remains the dominant external political and military actor in the Middle East.[175]

Although China's military presence and influence in the Middle East and North Africa is limited compared to the United States', it is possible that China may use its commercial energy ties to bolster its military relations in the region as China has done with its political relations. In this respect, the Congressional Research Service stated that:

> As in other regions, growing commercial ties facilitated the development of closer political relationships between China and many of its Middle Eastern counterparts.[176]

China may also use its energy diplomacy and relationships to bolster its arms relationships with the Middle East and North Africa. In this respect, "no Middle Eastern state has opposed China's entrance in the weapons sales area."[177]

Further, Alterman and Garver stated that:

> States as diverse as Israel and Iran have sought to develop arms relationships with China, driven by the direct benefits to their militaries as well as indirect political gains.[178]

If China is able to bolster its military relations and weapons sales to the Middle East and North Africa including Iran, it "has the potential to compete militarily with the United States and field disruptive technologies that could over time offset traditional U.S. military advantages absent U.S. counter-strategies."[179]

Further, China's actions could undermine international efforts to bring stability in the Middle East and North Africa in light of the fact that it views "their efforts as one of maintaining cooperative relations with all countries in the Middle East and North Africa."[180] Alterman and Garver opined that:

> The United States views such efforts as undermining international efforts to isolate troubling regimes and thus compel a change in their behavior.[181]

Moreover, in light of the fact that China is heavily dependent on crude oil imports to sustain its economy, it may use its weapons sales to the Middle East and North Africa to offset the cost of imported oil from the region. China may engage in a weapon for oil exchange to not only strengthen its overall ties but also reduce the cost of its imported crude oil. Erica Strecker Downs opined that:

> Militarily, it is possible that China could use its arms sales to the Middle East to foster ties to oil-producing nations and possibly to decrease its oil import bill.[182]

However, Alterman and Garver stated that "the U.S. role at this point dwarfs China's in supplying the Gulf States, although there remains potential for a greater Chinese role in the future.[183]

Therefore, it is imperative that the United States remain militarily engaged in the region not only for the sake of stability but also for India and Japan. Both India and Japan rely heavily on the region for oil. To have oil, the region must remain stable. Should the United States become less engaged, China may fill a military vacuum left vacant by the United States and possibly make the region less stable for India and Japan as well as other countries that depend on oil from the region.

China–Iran–India–Japan Military Relations

Indo-Iranian military relations are an important factor of the overall Indo-Iranian relationship.[184] India assisted Iran in upgrading its Russian-made military equipment in the 1990s and conducted joint naval maneuvers in 2003 and 2006.[185] India also has an intelligence station at their consulate in Zahedan, Iran, which would allow India to monitor Pakistan electronically.[186] Finally, India has assigned a military attaché to Iran.

Japanese–Iranian military relations are unknown except that Japan has stationed a defense attaché in Iran. Further, Japan has provided disaster relief from December 2003 to January 2004 when an earthquake hit southeastern Iran.

Sino-Iranian military relations are strong and are driven primarily by China's dependence on crude oil to sustain its economy and Iran's need for a powerful ally in the Middle East as well as a weapons supplier. China is Iran's largest supplier of conventional weapons and plays a major role in Iran's military modernization. China has provided to Iran anti-ship cruise missiles, surface-to-air missiles, combat aircraft, and fast attack patrol vessels, as well as advanced technology designed to expand the versatility of Iran's burgeoning cruise missile arsenal, which has "contributed significantly to what has become the central element of Iran's military rearmament—a revitalization of its naval forces."[187] Consequently, "U.S. intelligence agencies now estimate that Iran has the ability to shut off the flow of oil from the Persian Gulf for brief periods of time, even with a Western military presence in the region."[188]

China is also Iran's principal supplier of WMD technology and assistance including nuclear weapons according to the Pentagon. Preliminary liaisons began in the mid- to late-1980s and both countries are known to have signed nuclear accords in 1989 and 1991, paving the way for what would become a vibrant and multifaceted nuclear partnership.[189]

China's military technology transfers to Iran complicate U.S. Near East foreign policy to bring peace and stability to the region. A recent CRS report stated that "China's military cooperation with Iran has also proven to be a recurring problem in U.S.–China relations."[190] China's hesitance to sanction Iran in light of Iran's resistance to discontinue WMD development is an ongoing political crisis which invites a future military confrontation between Iran and the United States unless China decides to act prudently and take steps to de-militarize Iran of its nuclear technology and development.

If a military confrontation occurs between Iran and the United States, Iran could retaliate by mining or even attacking ships passing through the Strait of Hormuz. According to General Barry McCaffrey, Iran has three Kilo Class submarines, nineteen other mini-submarines, a significant sea mining capacity, shore based missile batteries, air defense capabilities, and a small number of high performance aircraft with missiles that could deny the two mile wide Strait of Hormuz sea transit zone to safe tanker traffic.[191] Iran could also place the Gulf Cooperation Council/Saudi Gulf oil terminals at risk.[192]

India and Japan's energy security would be jeopardized because crude oil shipments from the Gulf Region would cease or slowdown temporarily. Crude oil would have to be transported over land via pipeline to ports outside the Gulf Region. Fuel prices could rise along with other goods and services globally.

China–Kuwait–India–Japan Military Relations

Indo-Kuwaiti military relations are unknown. India and Kuwait do not have any bilateral military agreements, defense cooperation, or military exchanges. It is not known if India has cooperated with Kuwait militarily but India has assigned a military attaché to its embassy in Kuwait City.

Japanese–Kuwaiti military relations are also unknown. It is not known if Japan has cooperated with Kuwait other than posting a defense attaché at the Japanese embassy in Kuwait City and using Kuwait as a port of debarkation for international relief operations in Iraq in 2006 and 2009.

Sino-Kuwaiti military relations are developing. China and Kuwait have reciprocal military representation in each other's capitals and have conducted high-level military-to-military talks on future defense cooperation.

China's energy trading with Kuwait is very unlikely to result in either Sino-Indian or Sino-Japanese military competition for energy resources such as crude petroleum because Kuwait is a very minor oil exporter to China compared to India and Japan.

China–Saudi Arabia–India–Japan Military Relations

Indo-Saudi military relations are nascent at best.[193] There have been expressed interest in joint training, discussions on perspectives for peace and stability in the Middle East, naval cooperation in the Arabian Sea and Persian Gulf but no major progress has occurred thus far.[194] Despite the lack of progress, India has stationed a military attaché to Riyadh and conducted training ship visits in August and September 2011.

Japanese–Saudi military relations are unknown. It is not known if Japan has military representation in the Kingdom. However, the training squadron of Japan's Maritime Self-Defense Force (JMSDF) visited Jeddah in July 2012. Further, Japanese minesweepers made a goodwill port visit to King Abdulaziz Port in Dammam in October 2012.

Sino-Saudi military relations are developing in the wake of robust energy ties. China maintains a permanent defense attaché in the capital city of Riyadh and has conducted high-level talks with Saudi defense ministers. Further, China sold and transferred fifty Chinese immediate range ballistic missiles to Saudi Arabia in 1988. Apart from these actions, the extent of Sino-Saudi military relationship is unknown.

China's energy trading with Saudi Arabia is very unlikely to result in either Sino-Indian or Sino-Japanese military competition for energy resources such as crude petroleum because any such competition would invite U.S. military intervention. Further, the robust economic relationships China, India, and Japan enjoy with Saudi Arabia are too important to be negotiated militarily. In addition, China, India, and Japan do not have the capability to project military power in the Near East like the United States.

China–United Arab Emirates–India–Japan Military Relations

Indo-Emirati military relations are good. Both cooperate on counterterrorism and signed an agreement in 2003 which led to the creation of the India-UAE Joint Defense Cooperation Committee.[195] Additionally, their navies held joint exercises in 2003 and inaugurated high-level staff talks in January 2007.[196] Further, a small number of UAE military personnel have been attending training courses in India since 2003.[197]

Japanese–Emirati military relations are unknown. Japan and the United Arab Emirates do not have any bilateral military agreements, defense cooperation or military exchanges.

Sino-Emirati military relations are also unknown. However, China has a military attaché stationed in Abu Dhabi.

China's energy trading with the UAE is very unlikely to result in either Sino-Indian or Sino-Japanese military competition for energy resources such as crude petroleum because the UAE is a very minor oil supplier to China.

NOTES

1. Ashley J. Tellis, "China and India in Asia," in Francine R. Frankel and Harry Harding, eds., *The India and China Relationship: What the United States Needs To Know* (Washington, DC: Woodrow Wilson Press, 2004), 158.
2. Tellis, "China and India in Asia," 158.
3. Geoffrey Kemp, *The East Moves West: India, China, and Asia's Growing Presence in the Middle East* (Washington, DC: Brookings Institution Press, 2010), 119.
4. Yufeng Mao, "China's Interests and Strategy," in Joshua Eisenman, Eric Heginbotham and Derek Mitchell, eds., *China and the Developing World: Beijing's Strategy for the Twenty-First Century,* (Armonk, NY: Sharpe, 2007), 119.
5. Mao, "China's Interests and Strategy," 119.
6. Mao, "China's Interests and Strategy," 119.
7. Prince and Panda website, "PRINCE_SCORES.pdf," accessed October 17, 2013, http://www.princeandpanda.com.
8. Kemp, *East Moves West,* 23–24.
9. Kemp, *East Moves West,* 25.
10. Kemp, *East Moves West,* 25.
11. Kemp, *East Moves West,* 25.
12. Dawn Murphy, "China and the Middle East," testimony, June 6, 2013, before U.S.–China Economic and Security Review Commission.
13. Dawn Murphy, "China and the Middle East," testimony, June 6, 2013, before U.S.–China Economic and Security Review Commission.
14. Dawn Murphy, "China and the Middle East," testimony, June 6, 2013, before U.S.–China Economic and Security Review Commission.
15. Dawn Murphy, "China and the Middle East," testimony, June 6, 2013, before U.S.–China Economic and Security Review Commission.
16. Dawn Murphy, "China and the Middle East," testimony, June 6, 2013, before U.S.–China Economic and Security Review Commission.
17. Dawn Murphy, "China and the Middle East," testimony, June 6, 2013, before U.S.–China Economic and Security Review Commission.
18. Dawn Murphy, "China and the Middle East," testimony, June 6, 2013, before U.S.–China Economic and Security Review Commission.
19. Dawn Murphy, "China and the Middle East," testimony, June 6, 2013, before U.S.–China Economic and Security Review Commission.
20. Dawn Murphy, "China and the Middle East," testimony, June 6, 2013, before U.S.–China Economic and Security Review Commission.

21. Dawn Murphy, "China and the Middle East," testimony, June 6, 2013, before U.S.–China Economic and Security Review Commission.

22. Dawn Murphy, "China and the Middle East," testimony, June 6, 2013, before U.S.–China Economic and Security Review Commission.

23. Dawn Murphy, "China and the Middle East," testimony, June 6, 2013, before U.S.–China Economic and Security Review Commission.

24. Tellis, "China and India in Asia," 163.

25. Kemp, *East Moves West*, 52.

26. Kemp, *East Moves West*, 52.

27. Kemp, *East Moves West*, 52.

28. Kemp, *East Moves West*, 52.

29. Kemp, *East Moves West*, 52.

30. Ministry of Foreign Affairs of Japan, "Japan–Iran Relations," September 2012, accessed October 8, 2013, http://www.mofa.go.jp/region/middle_e/iran/index.html.

31. Kemp, *East Moves West*, 121.

32. Kemp, *East Moves West*, 121.

33. Kemp, *East Moves West*, 73.

34. Kemp, *East Moves West*, 75.

35. Ministry of External Affairs, Government of India, "India–Kuwait Relations," July 2013, accessed October 8, 2013, http://www.mea.gov.in/Portal/ForeignRelation/India-Kuwait_Relations.pdf .

36. Ministry of External Affairs, Government of India, "India–Kuwait Relations," July 2013, accessed October 8, 2013, http://www.mea.gov.in/Portal/ForeignRelation/India-Kuwait_Relations.pdf .

37. Ministry of Foreign Affairs of Japan, "Japan–Kuwait Relations," August 2013, accessed October 8, 2013, http://www.mofa.go.jp/region/middle_e/kuwait/index.html.

38. Embassy of the People's Republic of China in the State of Kuwait, "China and Kuwait," accessed October 8, 2013, http://kw.china-embassy.org/eng/sbgx/t580302.htm.

39. Ministry of Foreign Affairs of People's Republic of China, "China and Kuwait," accessed May 9, 2014, http://www.fmprc.gov.cn/mfa_eng/wjb_663304/zzjg_663340/xybfs_663590/gjlb_663594/2838_663666/.

40. Ministry of External Affairs, Government of India, "India–Saudi Arabia Relations," July 2013, accessed October 8, 2013, http://www.mea.gov.in/Portal/ForeignRelation/India-SaudiArabia_Relations.pdf.

41. Kemp, *East Moves West*, 37.

42. Kemp, *East Moves West*, 37–38.

43. Kemp, *East Moves West*, 38.

44. Kemp, *East Moves West*, 38.

45. Kemp, *East Moves West*, 38.

46. Kemp, *East Moves West*, 39.

47. Ministry of External Affairs, Government of India, "India–Saudi Arabia Relations," July 2013, accessed October 8, 2013, http://www.mea.gov.in/Portal/ForeignRelation/India-SaudiArabia_Relations.pdf.

48. Ministry of External Affairs, Government of India, "India–Saudi Arabia Relations," July 2013, accessed October 8, 2013, http://www.mea.gov.in/Portal/ForeignRelation/India-SaudiArabia_Relations.pdf.

49. Ministry of External Affairs, Government of India, "India–Saudi Arabia Relations," July 2013, accessed October 8, 2013, http://www.mea.gov.in/Portal/ForeignRelation/India-SaudiArabia_Relations.pdf.

50. Ministry of External Affairs, Government of India, "India–Saudi Arabia Relations," July 2013, accessed October 8, 2013, http://www.mea.gov.in/Portal/ForeignRelation/India-SaudiArabia_Relations.pdf.

51. Kemp, *East Moves West,* 124.

52. Kemp, *East Moves West,* 124.

53. Kemp, *East Moves West,* 124.

54. Ministry of Foreign Affairs of Japan, "Japan–Saudi Arabia Relations," January 2012, accessed October 8, 2013, http://www.mofa.go.jp/region/middle_e/saudi/index.html.

55. Kemp, *East Moves West,* 124.

56. Ministry of Foreign Affairs of People's Republic of China, "China–Saudi Arabia," 08/22/2011, accessed October 8, 2013, http://www.fmprc.gov.cn/eng/wjb/zzjg/xybfs/gjlb/2878/.

57. Jon B. Alterman and John W. Garver, *The Vital Triangle: China, The United States and the Middle East* (Washington, DC: CSIS Press, 2008), 33.

58. Alterman and Garver, *Vital Triangle,* 58.

59. Alterman and Garver, *Vital Triangle,* 59.

60. Jon B. Alterman, "China's Soft Power in the Middle East", in Carola McGiffert, ed., *Chinese Soft Power and Its Implications for the United States* (Washington, DC: CSIS Press, March 2009), 65.

61. Alterman and Garver, *Vital Triangle,* 35.

62. Ministry of Foreign Affairs of People's Republic of China, "China–Saudi Arabia," 08/22/2011, accessed October 8, 2013, http://www.fmprc.gov.cn/eng/wjb/zzjg/xybfs/gjlb/2878/.

63. Ministry of External Affairs, Government of India, "India–UAE Relations," August 2012, accessed October 8, 2013, http://www.mea.gov.in/Portal/ForeignRelation/uae-august-2012.pdf.

64. Kemp, *East Moves West,* 47.

65. Ministry of External Affairs, Government of India, "India–UAE Relations," August 2012, accessed October 8, 2013, http://www.mea.gov.in/Portal/ForeignRelation/uae-august-2012.pdf.

66. Ministry of External Affairs, Government of India, "India–UAE Relations," August 2012, accessed October 8, 2013, http://www.mea.gov.in/Portal/ForeignRelation/uae-august-2012.pdf.

67. Ministry of External Affairs, Government of India, "India–UAE Relations," August 2012, accessed October 8, 2013, http://www.mea.gov.in/Portal/ForeignRelation/uae-august-2012.pdf.

68. Ministry of Foreign Affairs of Japan, "Japan–United Arab Emirates Relations," April 2011, accessed October 8, 2013, http://www.mofa.go.jp/region/middle_e/uae/index.html.

69. Kemp, *East Moves West,* 126.

70. Kemp, *East Moves West,* 126.

71. Ministry of Foreign Affairs of Japan, "Japan–United Arab Emirates Relations," April 2011, accessed October 8, 2013, http://www.mofa.go.jp/region/middle_e/uae/index.html.

72. Kemp, *East Moves West,* 86.

73. Ministry of Foreign Affairs of People's Republic of China, "China–United Arab Emirates," 08/22/2011, accessed October 8, 2013, http://www.fmprc.gov.cn/eng/wjb/zzjg/xybfs/gjlb/2903/.

74. Mao, "China's Interests and Strategy," 118.

75. Dawn Murphy, "China and the Middle East," testimony, June 6, 2013, before U.S.–China Economic and Security Review Commission.

76. Dawn Murphy, "China and the Middle East," testimony, June 6, 2013, before U.S.–China Economic and Security Review Commission.

77. Dawn Murphy, "China and the Middle East," testimony, June 6, 2013, before U.S.–China Economic and Security Review Commission.

78. Dawn Murphy, "China and the Middle East," testimony, June 6, 2013, before U.S.–China Economic and Security Review Commission.

79. Dawn Murphy, "China and the Middle East," testimony, June 6, 2013, before U.S.–China Economic and Security Review Commission.

80. Dawn Murphy, "China and the Middle East," testimony, June 6, 2013, before U.S.–China Economic and Security Review Commission.

81. Kemp, *East Moves West,* 24.

82. Kemp, *East Moves West,* 24.

83. Kemp, *East Moves West,* 25.

84. Kemp, *East Moves West,* 25.

85. Kemp, *East Moves West,* 26.

86. Kemp, *East Moves West,* 128.

87. Kemp, *East Moves West,* 128.

88. Kemp, *East Moves West,* 13.

89. Kemp, *East Moves West,* 51.

90. Ministry of External Affairs, Government of India, "India–Iran Relations," October 2012, accessed October 8, 2013, http://www.mea.gov.in/Portal/ForeignRelation/India_Iran_Relations_-_October_2012.pdf.

91. Kemp, *East Moves West,* 54.

92. Ministry of External Affairs, Government of India, "India–Iran Relations," October 2012, accessed October 8, 2013, http://www.mea.gov.in/Portal/ForeignRelation/India_Iran_Relations_-_October_2012.pdf.

93. Ministry of External Affairs, Government of India, "India–Iran Relations," October 2012, accessed October 8, 2013, http://www.mea.gov.in/Portal/ForeignRelation/India_Iran_Relations_-_October_2012.pdf.

94. Ministry of External Affairs, Government of India, "India–Iran Relations," October 2012, accessed October 8, 2013, http://www.mea.gov.in/Portal/ForeignRelation/India_Iran_Relations_-_October_2012.pdf.

95. Kemp, *East Moves West,* 121.

96. Kemp, *East Moves West,* 121.

97. JETRO, "Japanese Trade and Investment Statistics," accessed October 8, 2013, http://www.jetro.go.jp/en/reports/statistics/.

98. JETRO, "Japanese Trade and Investment Statistics."
99. Ministry of Foreign Affairs of Japan, "Japan–Iran Relations," September 2012, accessed October 8, 2013, http://www.mofa.go.jp/region/middle_e/iran/index.html.
100. JAPEX, "Our Business: International Exploration and Production—Iran," accessed October 8, 2013, http://www.japex.co.jp/english/business/oversea/iran.html.
101. Kemp, *East Moves West,* 123.
102. Hisane Masaki, "Japan Energy: Goodbye Iran, Hello Iraq," *Asia Times Online,* November 7, 2006, accessed October 8, 2013, www.atimes.com/atimes/Japan/HK07Dh01.html.
103. "China Expects $40bln Trade Exchange with Iran in 2011," *Fars News Agency,* July 11, 2011; "China Keen to Expand Parliamentary Ties with Iran," *Fars News Agency,* April 19, 2011; and "Iran, China Sign 6 Industrial, Mineral Deals Worth $20B," *Mehr News Agency,* May 10, 2011.
104. Harold Scott and Alireza Nader, *China and Iran: Economic, Political and Military Relations* (Santa Monica, CA: RAND Corporation, 2012), 10.
105. Scott and Nader, *China and Iran,* 10.
106. Kemp, *East Moves West,* 77.
107. Kemp, *East Moves West,* 77.
108. Kemp, *East Moves West,* 77.
109. Kemp, *East Moves West,* 77.
110. Kemp, *East Moves West,* 77–78.
111. Ministry of External Affairs, Government of India, "India–Kuwait Relations," July 2013, accessed October 8, 2013, http://www.mea.gov.in/Portal/ForeignRelation/India-Kuwait_Relations.pdf.
112. Ministry of External Affairs, Government of India, "India–Kuwait Relations," July 2013, accessed October 8, 2013, http://www.mea.gov.in/Portal/ForeignRelation/India-Kuwait_Relations.pdf.
113. Ministry of External Affairs, Government of India, "India–Kuwait Relations," July 2013, accessed October 8, 2013, http://www.mea.gov.in/Portal/ForeignRelation/India-Kuwait_Relations.pdf.
114. Ministry of External Affairs, Government of India, "India–Kuwait Relations," July 2013, accessed October 8, 2013, http://www.mea.gov.in/Portal/ForeignRelation/India-Kuwait_Relations.pdf.
115. Ministry of External Affairs, Government of India, "India–Kuwait Relations," July 2013, accessed October 8, 2013, http://www.mea.gov.in/Portal/ForeignRelation/India-Kuwait_Relations.pdf.
116. Ministry of External Affairs, Government of India, "India–Kuwait Relations," July 2013, accessed October 8, 2013, http://www.mea.gov.in/Portal/ForeignRelation/India-Kuwait_Relations.pdf and Kemp, *East Moves West,* 51.
117. Ministry of External Affairs, Government of India, "India–Kuwait Relations," July 2013, accessed October 8, 2013, http://www.mea.gov.in/Portal/ForeignRelation/India-Kuwait_Relations.pdf.
118. Ministry of External Affairs, Government of India, "India–Kuwait Relations," July 2013, accessed October 8, 2013, http://www.mea.gov.in/Portal/ForeignRelation/India-Kuwait_Relations.pdf.

119. Ministry of Foreign Affairs of Japan, "Japan–Kuwait Relations," August 2013, accessed October 8, 2013, http://www.mofa.go.jp/region/middle_e/kuwait/index.html.

120. Ministry of Foreign Affairs of Japan, "Japan–Kuwait Relations," August 2013, accessed October 8, 2013, http://www.mofa.go.jp/region/middle_e/kuwait/index.html.

121. Kemp, *East Moves West,* 123.

122. Kemp, *East Moves West,* 123.

123. Khizar Niazi, "Kuwait Looks toward the East," Policy Brief, no. 26 (September 2009), accessed October 8, 2013, http://www.mei.edu/sites/default/files/publications/Niazi_0.pdf.

124. Niazi, "Kuwait Looks toward the East."

125. Niazi, "Kuwait Looks toward the East."

126. Niazi, "Kuwait Looks toward the East."

127. Kemp, *East Moves West,* 37.

128. Ministry of External Affairs, Government of India, "India–Saudi Arabia Relations," July 2013, accessed October 8, 2013, http://www.mea.gov.in/Portal/ForeignRelation/India-SaudiArabia_Relations.pdf.

129. "India–Saudi trade at $7 bn by 2010," *Rediff.com,* January 26, 2006, accessed October 9, 2013, http://in.rediff.com/money/2006/jan/23trade.htm.

130. Ministry of External Affairs, Government of India, "India–Saudi Arabia Relations," July 2013, accessed October 8, 2013, http://www.mea.gov.in/Portal/ForeignRelation/India-SaudiArabia_Relations.pdf.

131. Kemp, *East Moves West,* 39.

132. Ministry of External Affairs, Government of India, "India–Saudi Arabia Relations," July 2013, accessed October 8, 2013, http://www.mea.gov.in/Portal/ForeignRelation/India-SaudiArabia_Relations.pdf.

133. Kemp, *East Moves West,* 40.

134. Ministry of External Affairs, Government of India, "India–Saudi Arabia Relations," July 2013, accessed October 8, 2013, http://www.mea.gov.in/Portal/ForeignRelation/India-SaudiArabia_Relations.pdf.

135. Ministry of External Affairs, Government of India, "India–Saudi Arabia Relations," July 2013, accessed October 8, 2013, http://www.mea.gov.in/Portal/ForeignRelation/India-SaudiArabia_Relations.pdf.

136. Kemp, *East Moves West,* 125.

137. Kemp, *East Moves West,* 125.

138. Ministry of Foreign Affairs of Japan, "Japan–Saudi Arabia Relations," January 2012, accessed October 8, 2013, http://www.mofa.go.jp/region/middle_e/saudi/index.html.

139. Kemp, *East Moves West,* 125.

140. Ministry of Foreign Affairs of Japan, "Japan–Saudi Arabia Relations," January 2012, accessed October 8, 2013, http://www.mofa.go.jp/region/middle_e/saudi/index.html.

141. Kemp, *East Moves West,* 125.

142. Kemp, *East Moves West,* 125.

143. Dawn Murphy, "China and the Middle East," testimony, June 6, 2013, before U.S.–China Economic and Security Review Commission.

144. Ministry of Foreign Affairs of People's Republic of China, "China–Saudi Arabia," 08/22/2011, accessed October 8, 2013, http://www.fmprc.gov.cn/eng/wjb/zzjg/xybfs/gjlb/2878/.

145. Naser Al-Tamimi, "China–Saudi Relations: Booming Trade," *Al Arabiya News*, February 22, 2013, accessed October 8, 2013, http://english.alarabiya.net/views/2013/02/22/267670.html.

146. Al-Tamimi, "China–Saudi Relations: Booming Trade."

147. Kemp, *East Moves West*, 80.

148. Naser Al-Tamimi, "China–Saudi Relations: Booming Trade."

149. Kemp, *East Moves West*, 81.

150. Kemp, *East Moves West*, 81.

151. Kemp, *East Moves West*, 81.

152. Al-Tamimi, "China–Saudi Relations: Booming Trade."

153. Al-Tamimi, "China–Saudi Relations: Booming Trade."

154. Ministry of External Affairs, Government of India, "India–Saudi Arabia Relations," July 2013, accessed October 8, 2013, http://www.mea.gov.in/Portal/ForeignRelation/India-SaudiArabia_Relations.pdf.

155. Ministry of External Affairs, Government of India, "India–Saudi Arabia Relations," July 2013, accessed October 8, 2013, http://www.mea.gov.in/Portal/ForeignRelation/India-SaudiArabia_Relations.pdf.

156. Kemp, *East Moves West*, 47.

157. Ministry of External Affairs, Government of India, "India–UAE Relations," August 2012, accessed October 8, 2013, http://www.mea.gov.in/Portal/ForeignRelation/uae-august-2012.pdf.

158. Ministry of External Affairs, Government of India, "India–UAE Relations," August 2012, accessed October 8, 2013, http://www.mea.gov.in/Portal/ForeignRelation/uae-august-2012.pdf.

159. Ministry of External Affairs, Government of India, "India–UAE Relations," August 2012, accessed October 8, 2013, http://www.mea.gov.in/Portal/ForeignRelation/uae-august-2012.pdf.

160. Kemp, *East Moves West*, 48.

161. Ministry of External Affairs, Government of India, "India–UAE Relations," August 2012, accessed October 8, 2013, http://www.mea.gov.in/Portal/ForeignRelation/uae-august-2012.pdf.

162. Ministry of External Affairs, Government of India, "India–UAE Relations," August 2012, accessed October 8, 2013, http://www.mea.gov.in/Portal/ForeignRelation/uae-august-2012.pdf.

163. Kemp, *East Moves West*, 49.

164. Kemp, *East Moves West*, 126.

165. Kemp, *East Moves West*, 126.

166. Kemp, *East Moves West*, 126.

167. Ministry of Foreign Affairs of Japan, "Japan–United Arab Emirates Relations," April 2011, accessed October 8, 2013, http://www.mofa.go.jp/region/middle_e/uae/index.html.

168. Kemp, *East Moves West*, 126.

169. "China–Arab relations have centuries to come: UAE Minister," *People's Daily Online*, May 4, 2012, accessed October 8, 2013, http://english.peopledaily.com.cn/90883/7807397.html#.

170. Kemp, *East Moves West,* 87.

171. Dawn Murphy, "China and the Middle East," testimony, June 6, 2013, before U.S.–China Economic and Security Review Commission.

172. Kemp, *East Moves West,* 87.

173. Dawn Murphy, "China and the Middle East," testimony, June 6, 2013, before U.S.–China Economic and Security Review Commission.

174. Dawn Murphy, "China and the Middle East," testimony, June 6, 2013, before U.S.–China Economic and Security Review Commission.

175. Christopher M. Blanchard, et. al, *Comparing Global Influence: China's and U.S. Diplomacy, Foreign Aid, Trade, and Investment in the Developing World.* CRS Report RL34620 (Washington, DC: Library of Congress, August 15, 2008), 140.

176. Christopher M. Blanchard, et. al, *Comparing Global Influence,* 140.

177. Alterman and Garver, *Vital Triangle,* 66.

178. Alterman and Garver, *Vital Triangle,* 66.

179. Alterman and Garver, *Vital Triangle,* 96.

180. Alterman and Garver, *Vital Triangle,* 97.

181. Alterman and Garver, *Vital Triangle,* 97.

182. Erica Strecker Downs, *China's Quest for Energy Security* (Santa Monica, CA: RAND Corporation, 2000), 49.

183. Alterman and Garver, *Vital Triangle,* 73.

184. Kemp, *East Moves West,* 55.

185. Kemp, *East Moves West,* 55–56.

186. Kemp, *East Moves West,* 56.

187. Ilan Berman, "The Impact of Sino-Iranian Strategic Partnership," testimony, September 14, 2006, before U.S.–China Economic and Security Review Commission.

188. Berman, "The Impact of Sino-Iranian Strategic Partnership."

189. U.S. Department of Defense, *Proliferation: Threat and Response 1996* (Washington: U.S. Department of Defense, 1996), 14.

190. Blanchard, et. al, *Comparing Global Influence,* 149.

191. Barry McCaffrey, "Iran, Nukes, & Oil: The Gulf Confrontation" (Powerpoint Presentation Seminar for NBC Executives and Producers), January 12, 2012.

192. McCaffrey, "Iran, Nukes, & Oil: The Gulf Confrontation."

193. Kemp, *East Moves West,* 40.

194. Kemp, *East Moves West,* 40.

195. Kemp, *East Moves West,* 49.

196. Kemp, *East Moves West,* 49.

197. Kemp, *East Moves West,* 49.

6

Sub-Saharan Africa

In terms of real and potential energy producing states for China, India, and Japan, Sub-Saharan Africa (hereafter "Africa") is comprised of Angola, Equatorial Guinea, Gabon, Nigeria, and Sudan. For the purpose of discussion and analysis, the key oil producing states for China, India, and Japan are Angola, Nigeria and Sudan. South Sudan is also a key oil producing state because 75 percent of the oil production originates from the South, while the entire pipeline, refining, and export infrastructure is in the North.[1] This situation makes North Sudan and South Sudan interdependent and has caused contention between the two countries over pipeline and export transit fees.[2]

Although Africa has been less strategically important to China, India, and Japan in the past, many observers now see this region as an area of potential Sino-Indian and Sino-Japanese competition because of overlapping commercial interests. If either India or Japan seeks to diversify their oil supply sources and rely less on the Gulf States to supply the majority of their oil, Sub-Saharan Africa would be the next region for consideration.

Politically, India seeks to reinvent and rejuvenate its historic ties to Africa.[3] India is driven by energy security, the need for copper, uranium, and iron ore resources, trade and investment opportunities, and South-South cooperation.[4] Energy security is at the forefront of India's strategy in Africa as with China.[5] India is also driven by the need to keep up with Chinese investment in Africa because India has lost out to China in $12.5 billion worth of contracts in energy sector in 2009.[6] The April 2008 India-Africa Forum held in Delhi demonstrated India's commitment to securing its own footprint in Africa and competing with China.[7]

Politically, Japan seeks to contribute to the peace and stability of the region through nation building, development assistance, trade and investment, and response to global issues.[8]

In terms of peace and stability, Japan has dispatched Japanese Self-Defense Forces to Sudan and Somalia for peacekeeping operations (PKO) and election observation teams to support the democratization process in Niger, the Democratic Republic of the Congo and Liberia.[9]

In terms of development assistance, trade and investment, Japan has pledged to double its official development assistance (ODA) and private-sector investments as well as promote business in Africa.[10]

In terms of responding to global issues, Japan seeks the cooperation of African countries through United Nations Security Council reform and climate change.[11]

Politically, China seeks to increase its influence through international institutions. With respect to Africa, China has formalized its political influence through the Forum on China-Africa Cooperation (FOCAC). Established in 2000, FOCAC allows China to institutionalize its diplomatic overtures, technical training, debt relief, loans, grants and infrastructure projects and allows both China and the African nations to speak with one voice.[12] China also formalized its political influence though the United Nations in which it enjoys the support of many African nations.

China also seeks African support for the "one-China principle" in its efforts to isolate Taiwan from the international community of which China enjoys the support of the vast majority of Africa.[13] In this respect, only four African nations currently formally recognize Taiwan: Burkina Faso, Gambia, San Tome and Principe, and Swaziland.

Economically, China desires to strengthen its bilateral commercial ties with Africa. In terms of imports, China imports crude oil and minerals from Africa. In terms of exports, China seeks African markets for its manufactured goods.

However, commercial trade appears to be the dominant factor in Africa's overall relations with China. In this respect, the Congressional Research Service stated that:

> The dominant factor driving such ties is trade. Sino-African ties are underpinned by China's prodigious demand for Africa's plentiful commodities, notably oil and unprocessed metals and minerals, to supply its rapidly growing economy, and by African demand for Chinese goods and services.[14]

Further, Robert Rotberg opined that:

> China hardly wants to colonize, but it does have immense mercantilist ambitions. It ravenously seeks raw materials—petroleum, timber, ferrochrome, cobalt, platinum, copper, diamonds, and so on. Whatever primary resources are buried beneath the soils of Sub-Saharan Africa, China needs them to feed its massive industrial surge and ultimately America's consumer demand.[15]

Commercial energy trading is probably the important factor in Africa's overall relations with China, India, and Japan.

In terms of actual energy trading, crude oil and petroleum imports from Africa to India and Japan are 15 percent and 3 percent respectively according to FACTS Global Energy. Comparatively, China has imported more crude oil from Africa than any other region in the world except the Near East. Approximately 26 percent of China's crude oil is imported from Africa.

To achieve its political and economic objectives in Africa, China uses a "soft approach," agricultural and technical assistance, investment and economic support, military patronage and peacekeeping.

"China's soft approach refers to humanitarian, cultural and educational projects at promoting China's image and influence."[16]

Agricultural and technical assistance consists of agro-processing projects, and providing expertise and agricultural equipment, science and technology transfers, and technical training.

Investment and economic support is characterized by Chinese outward foreign direct investment (FDI) in infrastructure projects, grants, debt relief, and low-cost loans to African countries. It should be noted that loans are generally secured by oil. If an African nation defaults on a loan, it pays for it in crude oil and/or petroleum.

Military patronage consists of military cooperation, sales and assistance. In particular, military cooperation reinforces the burgeoning Sino-African strategic partnership because it helps to ensure the security of China's investments in Africa's natural resources and infrastructure.[17]

Chinese peacekeeping under a UN umbrella also helps China to achieve its political and economic objectives in Africa through legitimate means. Integrating Chinese forces through UN efforts helps to mitigate conflict in Africa and serve as an effective way to build military-to-military relations and political capital globally as well as local credibility in the African nation in question.[18]

Overall, China's economic engagement of Africa has the potential to intensify economic competition vis-à-vis India and Japan. However, political and military competition between China and India and China and Japan appears negligible, if any.

The PRINCE model predicts that China will continue to explore for oil in, and import oil from, Sub-Saharan Africa based on a probability of 93 percent or greater.[19]

SUB-SAHARAN AFRICA: POLITICAL IMPLICATIONS FOR INDIA AND JAPAN

In terms of diplomatic relations, China, India, and Japan generally have very good ties with Africa especially with those countries that produce crude petroleum. It is unlikely that China will be a political competitor of India and Japan in the short term because India and Japan's diplomatic presence is small compared to China. Further, India and Japan do not rely on Africa for the majority of their oil resources.

It is more likely that China will be a political competitor of the United States because China has the potential to undermine African and U.S. efforts to implement various good governance and anti-corruption initiatives in many developing African nations.[20] In this respect, through financial lending China can permit African countries to circumvent the conditions upon which assistance from the United States and international institutions such as the World Bank and the International Monetary Fund is predicted. Angola is a classic case and point where this practice has occurred.

Africa views China as a new market and a potential alternative to traditional U.S. dominance because of China's policy of internal non-interference.[21] After many decades of neglect and "underdevelopment" and being unable to benefit from the globalization process, many African countries have looked forward to new partners and new economic relations beyond their historic ties to Europe and the United States.[22] China fills this void because it is new, offers a fresh perspective to Africa's foreign relations, and demands nothing politically from the host country. The only exception is that any loans made from China to the host country may require collateral such as oil or the right to partially own or drill. China may also require that the host country trade with it exclusively.

China has created the Forum on China-Africa Cooperation (FOCAC) in Beijing in 2000 as a comprehensive effort to build mutually beneficial economic development, trade, and political relations with Africa rooted in principles of "South-South cooperation." Each FOCAC summit has produced a concrete action plan for Sino-African cooperation.

The 2000 summit held in Beijing "passed the Beijing Declaration of the Forum on China-Africa Cooperation and the Program for China-Africa Cooperation in Economic and Social Development."[23]

The 2003 meeting passed the Addis Ababa Action Plan (2004–2006) and consolidated trade relations, with China announcing the Special Preferential Tariff Treatment program, which removed import tariffs on 190 different items from twenty-five African countries.[24]

The 2006 summit adopted the Beijing Action Plan (BAP), which covered 2007 to the end of 2009. It was reportedly the largest international event held in China drawing China's top leaders and forty-eight high-level African government delegations, including forty-one heads of state.[25] Further, at the summit, President Hu Jintao made eight new major Chinese efforts to strengthen the Sino-African "strategic partnership" under FOCAC.[26] China's pragmatic policy focus on economic issues through increased political/diplomatic activity "has laid the foundation for the market entry of Chinese state-owned, as private, companies into Africa's economies" including China's oil companies CNPC, Sinopec, and CNOOC.[27]

In comparison to FOCAC, India has engaged Africa through the India-Africa Forum held in Delhi in April 2008. This was India's attempt to establish a footprint in Africa. China has also reached out diplomatically by maintaining an extensive network of diplomats in Africa.[28] China has embassies in all but four African countries. Further, frequent high-level visits by the Chinese president and premier have bolstered China's diplomatic presence. Moreover, reciprocal exchange visits by Chinese and African VIPs have resulted in bilateral cooperation agreements and often PRC state-financed business deals.[29]

China–Angola–India–Japan Diplomatic/Political Relations

Indo-Angolan diplomatic relations predate the pre-independence era of Angola when India supported Angolan independence from Portuguese colonial rule. Diplomatic relations are considered close and friendly.[30] Both are members of the Non-Aligned Movement and Angola supports India's candidature for a permanent seat in an expanded United Nations Security Council.[31]

Japanese–Angolan diplomatic relations are traditionally good and friendly since relations were first established in September 1976.[32] Since the establishment of peace on Angola in 2002, trade relations have become more important in Japanese–Angolan overall relations.[33]

Formal Sino-Angolan diplomatic relations were established in 1983 after China distanced itself from supporting the then National Union for the Total Independence of Angola (UNITA) rebels and recognized the ruling Popular Movement for the Liberation of Angola (MPLA) as the governing party.[34] Sino-Angolan diplomatic relations have improved since the end of the Cold War and are based on an emerging energy trading relationship as evidenced by the fact that Angola is China's largest supplier of petroleum in Africa at approximately 16 percent annually; that China offered Angola a $9 billion loan for infrastructure improvements in return for petroleum; and that China has burgeoning projects such as off-shore oil prospecting and exploration and refinery development. Further, Angola offers China a lighter or "sweeter" crude oil to import unlike Saudi Arabia's heavier or "sour" crude oil, which is more difficult to refine in terms of time and money.

China's energy trading with Angola is unlikely to result in either Sino-Indian or Sino-Japanese political competition for energy resources such as crude petroleum unless Angola becomes a major oil exporter to India and Japan. India and Japan respectively import from Angola approximately 11–12 percent and less than 1 percent annually. Also, India and Japan's diplomatic footprint is small compared to China.

China–Nigeria–India–Japan Diplomatic/Political Relations

Indo-Nigerian diplomatic relations are warm and friendly without any contentious issues.[35] Both are members of the Non-Aligned Movement and Nigeria is India's largest supplier of African crude oil at 11–12 percent annually.[36]

Japanese–Nigerian diplomatic relations are good and friendly since relations were first established in 1964. Like other African countries, trade has been an important element in overall relations between Japan and Nigeria.

Sino-Nigerian diplomatic relations are robust even though Nigeria is a very minor petroleum exporter to China. Diplomatic ties are largely driven by security cooperation and burgeoning bilateral trade, which is largely energy related such as some hydrocarbon trading and Nigerian oil and infrastructure development. It should be noted that Nigeria is becoming one of China's largest trading partners in Africa when it comes to hydrocarbons.

In terms of Nigeria's oil and infrastructure development, China has concluded a number of oil agreements between November 2004 and December 2006, in which Chinese companies obtained oil exploration licenses and equity (partial ownership) in Nigeria oil fields in exchange for investments in Nigeria's infrastructure totaling $3.1 billion or more. Further, in April 2006, President Abuja offered four exploration licenses to Chinese oil firms for investments in Nigeria's infrastructure totaling $4 billion.[37]

Although China's oil diplomacy has been generally favorable in Nigeria, it has been problematic. Chinese oil companies increasingly face armed reprisals from Nigerian militants of the Movement for Emancipation of Niger Delta (MEND) for the alleged Chinese expansion in the region or Chinese "colonialism." Many militants feel that this region has been exploited for the benefit of Nigerian elites and foreign oil companies to the detriment of the Nigerian people. MEND militants have responded accordingly by attacking oil installations and kidnapping foreign oil workers including Chinese nationals in early 2006, leading to a 20 percent reduction in Nigeria's oil production.[38] Consequently, Nigeria turned to China for military supplies to protect the oil fields.[39]

China's energy trading with Nigeria is unlikely to result in either Sino-Indian or Sino-Japanese political competition for energy resources such as crude petroleum because China imports less than 1 percent of crude oil annually from Nigeria. In comparison, India and Japan respectively imports 11–12 percent and less than 1 percent annually. Also, China's entry into the oil sector in Nigeria is fairly recent compared to India.[40]

China–Sudan–India–Japan Diplomatic/Political Relations

Indo-Sudanese diplomatic relations go back five thousand years. Bilateral relations are cordial and friendly.[41] Strong political relations continue between India and Sudan in different regional and international fora.[42] Both countries see their relations as a model of South-South cooperation, supporting the restructuring of the United Nations and the Non-Aligned Movement.[43]

Since diplomatic relations were first established there have been many bilateral agreements regarding oil and gas cooperation (India–Sudan Joint Working Group) and almost thirty bilateral agreements covering trade, science and technology, media, culture, consultations, business, small industries, legal matters, air services, and agriculture.[44] India also trains Sudanese diplomats at its Foreign Service Institute in New Delhi as well as its petroleum sector technical personnel.[45]

India and South Sudan have traditional and long standing ties before diplomatic relations were established on July 9, 2011, when South Sudan achieved independence.[46] India was the first country from Asia to open a consulate in Juba, the capital of South Sudan.[47] India was also one of the first countries to formally recognize South Sudan.[48] Since independence, many high-level bilateral visits have occurred to strengthen relations. India is committed to South Sudan's development in the areas of health, agriculture, irrigation, education, infrastructure, and energy.[49] Although Sudan and South Sudan are independent, South Sudan is especially important in terms of energy because 75 percent of Sudan's oil reserves are in South Sudan.[50]

Japanese–Sudanese diplomatic relations were established in January 1956.[51] Bilateral relations have been relatively good even though the two countries have engaged in little active trade and or the exchange of nationals.[52] Japan has been actively engaged in Sudan's development in terms of food security, primary health care, and water and sanitation since the 1980s to the early 1990s but security, political rights and the human rights situation have cast a dark shadow over their bilateral relationship.[53]

Japanese–South Sudanese diplomatic relations were established on July 9, 2011.[54] Little is known about the bilateral relationship except that Japan is providing technical assistance, infrastructure development, the deployment of an engineering contingent of the United Nations Mission in South Sudan (UNMISS), and constructing a new $100 million bridge over the Nile river in Juba.[55] Moreover, the Toyota Tsusho Corporation agreed to build an oil pipeline from South Sudan to Lamu Port in Kenya.[56] Further, South Sudan signed a memorandum of understanding (MOU) with Kenya, Ethiopia, and Djibouti to possibly build a pipeline either to the Kenyan Port of Lamu or to the Port of Djibouti through Ethiopia.[57] If the pipeline is built, South Sudan will no longer be economically hostage to Sudan especially since oil represents 98 percent of South Sudan's total government revenues.[58]

Sino-Sudanese diplomatic relations were established in January 1959. Since diplomatic ties were established, China and Sudan have enjoyed a traditional friendship.[59]

Their political relations are considered close and firmly bilateral.[60] They extend to forms of cooperative interaction in regional organizations like FOCAC or the China-Arab Cooperation Forum and international organizations like the United Nations or the G77.[61]

Sino-South Sudan diplomatic relations were established when South Sudan achieved independence on July 9, 2011. Very little is known about the relationship except that China does not interfere in South Sudan's domestic politics. Also, South Sudan's political relations with China are recent and thus much less developed.[62]

China's energy trading with Sudan and South Sudan is unlikely to result in political competition for oil resources between China and India or China and Japan because India and Japan are not heavily dependent on Sudan and South Sudan's oil. Further, the political situation is delicate given that Sudan and South Sudan are interdependent economically. Their economies and government revenues are almost entirely based on oil. For Sudan, oil represents over half their state revenue and 90 percent of export earnings.[63] For South Sudan, oil represents almost 100 percent of state revenues.[64]

SUB-SAHARAN AFRICA: ECONOMIC IMPLICATIONS FOR INDIA AND JAPAN

China's interest in Africa is primarily economic: energy trading, energy prospecting, and the trade of goods and services. In terms of energy trading, Africa is China's second-largest crude oil and petroleum supplier after the Near East. Moreover, in terms of oil prospecting, China has ongoing upstream sector activities in Angola, Chad, Congo-Brazzaville, Cote d'Ivoire, Equatorial Guinea, Gabon, Mauritania, Nigeria, Sudan, and Uganda and prospective upstream sector developments in Kenya, Madagascar, Mali, and Niger. See appendix B. Finally, in terms of goods and services, Africa is a major import market for Chinese raw and manufactured goods such as textiles, apparel, footwear, machinery and transportation equipment, ores and metals, agricultural raw materials, manufactured materials, and processed food and beverages.

China's principal strategic organ for developing and sustaining its economic relations (as well as its political relations) with Africa is the Forum on China-Africa Cooperation (FOCAC). Created in Beijing in 2000, the FOCAC was instrumental to not only advancing China's economic (and political) interests in Africa but also cementing its economic (and political) influence. FOCAC also ushered a new era of trade cooperation between China and Africa and made China Africa's second-largest trading partner after the United States.[65]

Through FOCAC China was able to spawn further economic development and influence in Africa. At the 2006 FOCAC summit, China initiated the creation of Special Economic Zones (SEZs), which will provide liberalized investment environments focused on strategic industries to attract Chinese companies.[66] In this respect, Chinese companies have heavily increased investments in Africa particularly in the oil sector in Angola, Nigeria, and Sudan.[67] Nigeria is particularly important because it is a growing investment destination for Chinese oil companies like Sinopec, PetroChina (a CNPC subsidiary), and CNOOC.

China has also used the Export-Import (EXIM) Bank for economic development and cooperation in Africa. Founded in 1994, the Export-Import Bank of China is the sole state-owned entity of China to dispense economic aid worldwide. According to the World Bank, the Export-Import Bank of China has doled out over $12.5 billion in infrastructure development.[68] Aid is mainly concentrated in oil rich countries, such as Angola, Mozambique, Nigeria, Sudan, and Zimbabwe.[69]

India and Japan's interest in Africa is also economic. Both countries are eager to tap into Africa's rich oil resources and diversify their sources of crude oil and gas supplies away from the volatile Middle East region.[70] For Japan, the demand for fossil fuels has been even greater in the wake of the 2011 Fukushima nuclear disaster. Japan has been seeking alternate energy sources to generate electricity nationally. Fossil fuels like petroleum and natural gas may be a safe alternate fuel source to nuclear power.

Africa also serves as an increasing destination for Indian and Japanese exports even though the volume of trade is low in comparison to China. Indian exports to Africa consist mainly of manufactured items, chemical products, machinery, and transport equipment.[71] For Japan, it is motor vehicles, machinery and equipment, and steel.

India's trade relationship with Africa has been promoted through political and economic initiatives. Politically, India created three joint secretaries to manage three regional divisions.[72] Economically, the India-Africa Partnership Project and Focus Africa Program were launched through the Export-Import (EXIM) Bank of India to deepen economic ties with the continent.[73]

Japan's approach to bilateral trade has been much more subtle. In this respect, Japan has fostered trade relationships as well as diplomatic ones through the Tokyo International Conference on African Development

(TICAD). TICAD was initiated and created by the Japanese Government after the collapse of the Soviet Union and its Socialist Allies.[74] TICAD is essentially an unavoidable approach or mechanism of Japanese foreign policy toward Africa.[75] Through TICAD, Japan provided large scale loans and financial assistance to African countries with the aim of not only securing its bid for permanent membership at the United Nations Security Council but also securing a stake in Africa's rich oil resources.[76]

Both India and Japan will have to compete economically with China for crude oil energy sources. Competition will be keen given China has already established itself both politically and economically. In this respect, India and Japan have to match China's adroit use of soft power instruments such as economic aid and debt relief incentives.

China–Angola–India–Japan Economic Relations

Indo-Angolan economic relations are traditional and friendly. Although India and Angola have not signed any bilateral agreements for joint oil exploration and production, decisions have been made in principle for the protection and promotion of investments in Angola. In this respect, two memorandums of understandings (MOUs) were signed in 2010 which envision bilateral collaboration in Angola, India, and other countries.[77]

Angola figures prominently in terms of crude oil exports to India. Angola is India's second-largest source of crude oil after Nigeria.[78] In 2011, 11 percent of Angola's oil exports went to India.[79] Further, approximately 4 percent of India's annual crude oil imports originated from Angola in 2012 according to FACTS Global Energy.

In terms of overall trade, Indo-Angolan trade was $8.7 billion in 2012–2013.[80] In fact, India is Angola's second-largest trading partner after China, sharing about 10.6 percent of Angola's external trade.[81] While Angola exports crude oil and diamonds to India, India exports to Angola tractors and transport vehicles, agricultural machinery and implements, food and meat products, pharmaceuticals and cosmetics, tea, basmati rice, spirits and beverages, finished leather, and paper/wood products.[82] India has also extended lines of credit to Angola which has been utilized in sectors such as Railway Rehabilitation Project, the supply of agricultural equipment and Indian tractors.[83]

Japanese–Angolan economic relations are amicable. Bilateral trade is important to both countries even though Japan has very small energy stakes in Angola and imports less than 1 percent of its oil annually from Angola. In terms of energy stakes, Japan's national oil company AJOCO has three stakes in shallow-water oil blocks (20 percent in Blocks 3/05 and 3/05–A and 12.5 percent in Block 3/85–91), where production is declining.[84]

Angola could play a larger role in Japan's energy security in the future if Japan relies more on African oil.

In terms of bilateral trade, Japanese–Angolan trade was almost $300 million in 2006.[85] While Angola exports crude petroleum, fish and shellfish, and animal food to Japan, Japan exports to Angola automobiles, machinery parts, and steel products.[86] Japan has also two companies operating in Angola.[87]

Sino-Angolan oil economic relations are robust and strategic as evidenced by the fact that Angola is China's largest petroleum energy trading partner in Africa. In this respect, Angola exports about 15 percent of China's total annual petroleum and China has oil equity stakes, offshore oil prospecting ventures, an oil refinery project under development and oil for credit deals of $2 billion. This $2 billion aid package funded through China's EXIM Bank includes funds for Chinese companies to build railroads, schools, roads, hospitals, bridges, and offices; lay a fiber-optic network, and train Angolan telecommunications workers.[88]

In terms of bilateral trade, Sino-Angolan trade was $24.8 billion in 2010.[89] In fact, Angola is China's second largest trading partner in Africa after South Africa.

Angola exports to China crude petroleum, diamonds, granite, porphyry, and basalt while China principally exports to Angola steel and iron bars, batteries, cement, and automobiles.[90,91]

China's energy trading with and oil stakes in Angola will result in economic competition between China and India and China and Japan. India has already been outmaneuvered by China when the Chinese oil company Sinopec sidelined India's state-owned Oil and Natural Gas Corporation (ONGC) for 50 percent ownership of British Petroleum's (BP) operated Block 18.[92] ONGC's $310 million offer for infrastructure development could not compete with Sinopec's offer of $725 million.[93] China has also pledged a $2 billion aid package to Angola. Sinopec's offer of $725 million coincided with the announcement of China's EXIM Bank loan of $2 billion to finance infrastructure reconstruction.[94] Further, between 2004 and 2010, China's EXIM Bank has extended a total of $ 10.5 billion in oil-backed credit lines to the Angolan Government.[95] Angola initially approached Japan in 2003 for reconstruction funding but Japan demanded from Angola to improve its relations with the International Monetary Fund (IMF).[96]

China has carved out a prominent position in Angola's economy as its largest partner, a major provider of funds and the largest operator in Angola's infrastructure project.[97] China's effective use of soft power instruments such as capital investment makes it a fierce competitor of India and Japan for crude oil resources.

China–Nigeria–India–Japan Economic Relations

Indo-Nigerian economic relations are robust as evidenced by the fact that India is Nigeria's second-largest trading partner.[98] India is also Nigeria's largest importer of crude oil importing 8–12 percent of its oil requirements annually.[99] India also has energy stakes in Nigeria. See appendix C.

Nigeria is India's largest trading partner in Africa. Nigeria is also the largest market in Africa for Indian imports.[100]

In terms of bilateral trade, Indo-Nigerian trade was $16.5 billion between April 2012 and March 2013. While Nigeria principally exports crude oil to India, India principally exports to Nigeria refined petroleum products, machinery and instruments, pharmaceuticals, electronics, steel, power transmission equipment, and transport equipment.[101] India also has over one hundred companies operating in Nigeria in the pharmaceutical, steel, and power transmission sectors.[102]

Japanese–Nigerian economic relations are not as robust as Indian trade ties with Nigeria. Yet, bilateral trade is important to both countries even though Japan imports less than 1 percent of its oil annually from Nigeria according to FACTS Global Energy. Like Angola, Nigeria could play a larger role in Japan's energy security in the future if Japan relies more on African oil and less on Gulf State oil.

In terms of bilateral trade, Japanese–Nigerian trade was almost $1.13 billion in 2009.[103] While Nigeria exports crude petroleum and liquefied natural gas to Japan, Japan exports to Nigeria automobiles and steel products.[104]

Sino-Nigerian economic relations are growing by virtue of the fact that Nigeria has been designated by China as a Special Economic Zone (SEZ), especially for Chinese oil companies. Bilateral trade has developed slowly since the mid-1990s but has reached $13 billion in 2013.[105] In terms of bilateral trade, Nigeria exports cocoa beans, rubber, cashew nuts, hide and skin, agricultural products, and crude petroleum to China while China exports to Nigeria motorcycles, machinery equipment, auto parts, rubber tires, chemical products, textiles and garments, footwear, and cement.[106]

China seeks closer economic ties with Nigeria considering Nigeria is Africa's largest supplier of crude oil.[107] China has invested nearly $13 billion in Nigeria for infrastructure development in critical sectors such as roads, the railways, bridges, airports, and public transportation.[108] China's generous economic aid package to Nigeria is tied to its ongoing investments in Nigeria's petroleum/mining sector, which includes equity stakes and oil exploration and development. See appendix B. Despite China's energy stakes, it imports less than 1 percent of its crude oil annually from Nigeria.

China's burgeoning investment in Nigeria's petroleum/mining sector, in which trade-related finance is tied to another major area of investment, infrastructure construction, is a matter of economic concern for India and Japan because China may potentially outdo Indian and Japanese competition by paying above-market prices to guarantee access to oil supplies. For example, China paid Nigeria $3.1 billion between 2004 and 2006 in infrastructure capital in return for offshore drilling rights. China also paid Nigeria $4 billion in April 2006 in return for four oil exploration licenses. In addition, China has invested nearly $13 billion in infrastructure development in return for energy stakes to date. Because Nigeria is heavily dependent on foreign direct investment to sustain its petroleum/mining-based economy and China has been more willing to make sweeter infrastructure deals in exchange for drilling rights, Nigeria may choose to work with Chinese oil companies over other oil companies in the future. These companies may include Indian national oil companies or Japanese independent oil companies. The prospect of economic favoritism does not bode well for China's competition. India and Japan will most likely economically compete with China for oil energy resources if China continues to outspend its competitors on foreign aid.

China–Sudan–India–Japan Economic Relations

Indo-Sudanese economic relations appear to be strong. Bilateral trade was $888.28 million in 2012–2013.[109] In terms of bilateral trade, Sudan exports to India minerals and fuels, raw hides and skins, agricultural products, leather, and metals and ores while India exports to Sudan sugar and confectioneries, chemicals and pharmaceuticals, machinery and equipment, transport equipment, manufactured goods, textiles and apparels, and foodstuffs.[110]

India has oil stakes in Sudan. See appendix C. Further, India jointly operates the Greater Nile Petroleum Operating Company (GNPOC) with China, Malaysia and Sudan. India also imports approximately less than 1 percent of its crude oil annually from Sudan. Four percent of Sudanese oil exports were destined for India in 2011.[111] This is half of Sudan's crude oil exports to Japan in the same year.[112]

Indo-South Sudanese economic relations have been developing since diplomatic relations were established in 2011. A number of Indian companies have expressed interest in doing business in South Sudan.[113]

Japanese–Sudanese economic relations appear to be robust. Bilateral trade was $4.3 billion in 2008.[114] In terms of bilateral trade, Sudan exports to Japan crude petroleum and petroleum products, gum arabic, sesame seeds and cotton while Japan exports to Sudan machinery, industrial products, and transportation equipment.[115]

Although Japan does not have oil stakes in Sudan, it does import crude oil. Japan imports less than 2 percent of its crude oil annually from Sudan. Approximately 8 percent of Sudanese crude oil exports were destined for Japan in 2011.[116] This was double the percentage of exports to India for the same year.

Japanese–South Sudanese economic relations are developing. In terms, of energy trading, South Sudanese oil is sought out by Asian importers like Japan because of its low sulfur and high waxy content.[117] Further, power shortages in Japan in wake of the Fukushima incident have heightened the demand for South Sudan's two crude oil export grades: the Dar Blend and Nile Blend.[118] Thus, South Sudan figures prominently in Japan's energy security.

Sino-Sudanese economic relations are robust as evidenced by the fact China is Sudan's largest trading partner. Bilateral trade was $8.6 billion in 2010.[119] In terms of bilateral trade, Sudan exports to China crude petroleum, cotton, gum arabic, and oil seeds while China exports to Sudan textiles, building materials, light machinery and transport equipment.[120]

Sino-Sudanese oil economic relations are also strong as evidenced by the fact that Sudan exports approximately 6 percent of China's annual oil requirements and China has high hydrocarbon energy stakes in Sudan in terms of equity shares in oil fields, prospecting and oil pipeline construction. See appendix B. Chinese investment totaling $2.3 billion in Sudan's oil industry has resulted in 936 miles of oil pipelines, the construction of a crude oil processing plant with a capacity of 18.3 million barrels per year and several gas stations.[121] With Chinese financial and technical assistance, Sudan established its own oil industry, consisting of prospecting, exploitation, refining and transportation facilities, and sales networks.[122] China has helped Sudan transform itself from an oil importing country into an oil exporting country.[123]

Sino-South Sudanese economic relations have been developing since South Sudan became an independent country in 2011.[124] China and South Sudan have signed bilateral trade, economic, and technical agreements and established a bilateral joint economic and trade committee.[125] Further, sixty Chinese-funded enterprises have registered in South Sudan, related mainly to petroleum, construction, road and bridge, and communications as of October 2012.[126]

India and Japan will experience economic competition from China given China's substantial energy stakes and investment in Sudan. Although India jointly operates GNPOC with China, China's investment portfolio is much stronger than India.

The construction of an estimated $4 billion oil pipeline from South Sudan to the Kenyan port of Lamu by the Toyota Tsusho Corporation could boost South Sudan's oil exports given South Sudan is landlocked and relies solely on transport facilities that pass through neighboring Sudan.[127] Pipeline construction and operation could enhance not only Japanese–South Sudanese diplomatic relations but also trade ties. Japan would have some political and economic leverage considering 75 percent of the oil reserves are in the South Sudan. Beyond Japan, an oil pipeline from South Sudan to the Port of Djibouti via Ethiopia would also reduce South Sudan's reliance on Sudan.

SUB-SAHARAN AFRICA: MILITARY IMPLICATIONS FOR INDIA AND JAPAN

China's demand for African oil is unlikely to have adverse military consequences for either India or Japan or lead to a military competition for strategic energy resources in the short or medium-term because Africa is not a critical part of China's national security. "Africa is a security issue only in terms of China's efforts to secure the energy, mineral and timber resources to fuel its economy."[128] Further, Chinese military activities (military assistance, arms transfers, and peacekeeping operations) are essentially a function of China's much greater use of soft power in advancing its greater diplomatic, economic, and security interests. However, if the oil producing regions become politically unstable to the point where indigenous security forces cannot protect oil fields and foreign oil workers, then China may step-in and provide military training, assistance, personnel and equipment to protect its strategic oil interests. Angola, Cote d'Ivoire, Nigeria, and Sudan are the best examples in recent times where China has provided either military assistance or equipment to protect its oil interests. To Angola and Nigeria, China provided military training and weapons; to Cote d'Ivoire and Sudan, China provided peacekeepers respectfully under the UN umbrella. It should be noted that China's most significant peacekeeping involvements in Africa are linked to countries where it has either a major interest in natural resources such as oil or there are special considerations.[129] Therefore, it appears that Chinese military activities in Africa are integral and subservient to China's larger national goals of securing vital energy resources rather than as an independent political actor.

China historically has had good military relations with Africa being a major supplier of small arms and light weapons. However, China cannot project military power globally; it doesn't have operational experience. Further, China doesn't have military bases in Africa; it doesn't seek traditional alliances as a matter of national policy.[130] Finally, China's military efforts are a corollary to Beijing's much greater use of soft power in advancing its strategic interests. Therefore, it is unlikely that there will be a military competition for strategic energy resources between China and India or China and Japan in the short or medium-term even though China's military presence and role in Africa is growing.[131]

In terms of military presence and influence, China has fourteen of 107 worldwide military attaché offices in Africa, of which two host countries have proven oil reserves: Nigeria and Sudan. Further, between 2001 and 2006, China has conducted thirty major military exchanges and 110 bilateral security-related meetings and consultations; is currently participating in six UN peacekeeping operations in Western Sahara (MINURSO), Darfur (UNAMID), Liberia (UNMIL), Cote d'Ivoire (UNOCI), the Democratic Republic of Congo (MONUSCO), and Abyei (UNISFA) as September 2013; and supplied arms to at least twenty-one African countries of which six have real or potential oil reserves: Gabon, Kenya, Mali, Mauritania, Nigeria and Sudan according to Stockholm International Peace Research Institute (SIPRI).[132] Whether China's growing military presence and influence will result in a military competition for strategic energy resources in the long-term remains to be seen. It all depends on whether China will resort to military force to accomplish its national political-economic goals.

India and Japan's military engagement of Africa has been very minimal and has been limited to UN peacekeeping operations. Japan has UN peacekeepers in Abyei (UNISFA) while India has them in Western Sahara (MINURSO), Liberia (UNMIL), Cote d'Ivoire (UNOCI), the Democratic Republic of Congo (MONUSCO), South Sudan (UNMISS) and Abyei (UNISFA).

India has supplied maritime patrol aircraft to Mauritius, light helicopters to Namibia and Suriname, maritime patrol aircraft to the Seychelles, and offshore patrol vessels to Nigeria according to SIPRI.[133] Japan has not supplied weapons or munitions to Africa but has constructed and opened the Djibouti naval base in July 2011. The base will provide Japan with a significant boost to its antipiracy efforts in the Red Sea, while helping to ensure the security of its shipping through the Indian Ocean's SLOCs.[134] Further, the new base will also play a major role in increasing Japan's strategic influence in the region.[135]

China–Angola–India–Japan Military Relations

Indo-Angolan military relations are unknown. According to the Indian Ministry of Defence, India and Angola do have any bilateral military agreements, defense cooperation, or military exchanges. India appears to have supported Angola in the past militarily through the UN peacekeeping forces during the Angolan Civil War between the Popular Movement for the Liberation of Angola (MPLA) and the National Union for Total Independence of Angola (UNITA) in the 1990s.

Japanese–Angolan military relations are unknown. According to the Japanese Ministry of Defense, Japan and Angola do not have any bilateral military agreements, defense cooperation, or military exchanges.

Sino-Angolan military relations are growing. Angola is becoming a more important partner for military cooperation with China because Angola is China's primary source of oil in Africa.[136] Since 2000, China and Angola have conducted several high-level military visits. Further, in 2007, China has provided military technical assistance and supplied military equipment to the Angolan armed forces. In addition, Angola military personnel have received training in China.[137]

China's energy trading in Angola is unlikely to result in a military competition for oil resources between China and India or China and Japan. Japan imports less than 1 percent of its oil requirements annually from Angola.

China–Nigeria–India–Japan Military Relations

Indo-Nigerian military relations are developing. According to the Indian Ministry of Defence, India and Nigeria do not have any bilateral military agreements. However, India's army chief of staff visited Nigeria in November 2005 and pledged to help train and modernize the Nigerian military.[138] India will supply Nigeria with offshore patrol vessels in 2015 according to SIPRI.[139]

Japanese–Nigerian military relations are also unknown. According to the Japanese Ministry of Defense, Japan and Nigeria do not have any bilateral military agreements, defense cooperation, or military exchanges.

Sino-Nigerian military relations are good even though Nigeria is a minor exporter of petroleum to China. Both countries have reciprocal defense attaché offices in their respective capitals. Further, both have conducted high-level military visits and meetings between 2000 and 2006. In addition, China has supplied Nigeria with small arms in 1997 and 1998, fighter aircraft, and air-to-air missiles in 2005 and licensed offshore patrol vessels in 2012 according to SIPRI.[140]

China's energy trading in Nigeria is unlikely to result in a military competition for oil resources between China and India or China and Japan. China and Japan import less than 1 percent of their oil requirements annually from Nigeria.

China–Sudan–India–Japan Military Relations

Indo-Sudanese military relations are unknown. According to the Indian Ministry of Defence, India and Sudan do not have any bilateral military agreements, defense cooperation, or military exchanges.

Japanese–Sudanese military relations are also unknown. According to the Japanese Ministry of Defense, Japan and Sudan do not have any bilateral military agreements, defense cooperation, or military exchanges.

Sino-Sudanese military relations are robust as are their oil energy relations. Both countries have reciprocal defense attaché offices in their respective capitals. Further, both have conducted high-level military visits and meetings between 2000 and 2006. In addition, China has supplied Sudan with small arms and fighter aircraft between 1997 and 2004, tanks between 2002 and 2003, portable surface-to-air missiles in 2005 and licensed infantry fighting vehicles in 2003. Finally, China has been engaged in UN peacekeeping operations (UNAMID) in Darfur.

Indo-South Sudanese military relations are unknown. According to the Indian Ministry of Defence, India and South Sudan do not have any bilateral military agreements, defense cooperation or military exchanges.

Japanese–South Sudanese military relations are also unknown. According to the Japanese Ministry of Defense, Japan and South Sudan do not have any bilateral military agreements, defense cooperation, or military exchanges.

Like India and Japan, Sino-South Sudanese military relations are unknown. China and South Sudan do not appear to have any bilateral military agreements, defense cooperation, or military exchanges.

However, China, India, and Japan have been engaged in UN peacekeeping operations (UNMISS) in South Sudan.

China's energy trading with South Sudan is unlikely to result in military competition for oil resources between China and India and China and Japan.

NOTES

1. U.S. Energy Information Administration, U.S. Department of Energy, "Country Analysis Briefs: Sudan and South Sudan," March 19, 2012, accessed October 11, 2013, http://www.eia.gov/countries/analysisbriefs/cabs/Sudan/pdf.pdf.

2. U.S. Energy Information Administration, U.S. Department of Energy, "Country Analysis Briefs: Sudan and South Sudan," March 19, 2012, accessed October 11, 2013, http://www.eia.gov/countries/analysisbriefs/cabs/Sudan/pdf.pdf.

3. Sanusha Naidu, "India's African Relations: In the Shadow of China?" in Fantu Cheru and Cyril Obi, eds., *The Rise of China and India in Africa: Challenges, Opportunities and Critical Interventions* (New York, NY: Zed Books, 2010), 35.

4. Abdullah Verachia, "Unpacking India's African Engagement" (Powerpoint presentation before the Frontier Advisory, Paris, France), October 2010, 24–26.

5. Cyril Obi, "African Oil in the Energy Security Calculations of China and India" in Fantu Cheru and Cyril Obi, eds., *The Rise of China and India in Africa:*

Challenges, Opportunities and Critical Interventions (New York, NY: Zed Books, 2010), 187.

6. Abdullah Verachia, "Unpacking India's African Engagement," 27.

7. Fantu Cheru and Cyril Obi, "Introduction—Africa in the Twenty-First Century: Strategic and Development Challenges," in Fantu Cheru and Cyril Obi, eds., *The Rise of China and India in Africa: Challenges, Opportunities and Critical Interventions* (New York, NY: Zed Books, 2010), 5.

8. Ministry of Foreign Affairs, Japan, "Japan's Foreign Policy by Region," in *Diplomatic Bluebook 2012* (Tokyo, Japan), April 2012, 29.

9. Ministry of Foreign Affairs, Japan, "Japan's Foreign Policy by Region," in *Diplomatic Bluebook 2012* (Tokyo, Japan), April 2012, 29.

10. Ministry of Foreign Affairs, Japan, "Japan's Foreign Policy by Region," in *Diplomatic Bluebook 2012* (Tokyo, Japan), April 2012, 29.

11. Ministry of Foreign Affairs, Japan, "Japan's Foreign Policy by Region," in *Diplomatic Bluebook 2012* (Tokyo, Japan), April 2012, 29.

12. Joshua Eisenman, "China's Post-Cold War Strategy in Africa: Examining Beijing's Methods and Objectives," in Joshua Eisenman, Eric Heginbotham, and Derek Mitchell, eds., *China and the Developing World: Beijing's Strategy for the Twenty-First Century* (Armonk, New York: M.E. Sharpe, 2007), 35.

13. Eisenman, "China's Post-Cold War Strategy in Africa," 36.

14. Christopher M. Blanchard, et. al, *Comparing Global Influence: China's and U.S. Diplomacy, Foreign Aid, Trade, and Investment in the Developing World.* CRS Report RL34620 (Washington, DC: Library of Congress, August 15, 2008), 114.

15. Robert I. Rotberg, "China's Quest for Resources, Opportunities, and Influence in Africa" in Robert I. Rotberg, ed., *China Into Africa: Trade, Aid and Influence.* (Washington, DC: Brookings Institution Press, 2008), 1.

16. Eisenman, "China's Post-Cold War Strategy in Africa," 43.

17. J. Peter Pham, "China's African Strategy and Its Implications for U.S. Interests," *American Foreign Policy Interests,* Issue 28, 2006, 247.

18. Eisenman, "China's Post-Cold War Strategy in Africa," 50.

19. Prince and Panda website, "PRINCE_SCORES.pdf," accessed October 17, 2013, http://www.princeandpanda.com.

20. U.S. Library of Congress, Congressional Research Service, *China's Foreign Policy and "Soft Power" in South America, Asia, and Africa: A Study Prepared for the Committee on Foreign Relations United States Senate* (Washington, DC: U.S. Government Printing Office, 2008), 128 and Blanchard, et. al, *Comparing Global Influence,* 117–118.

21. Wenran Jiang, "China's Emerging Strategic Partnerships in Africa" in Robert I. Rotberg, ed., *China Into Africa: Trade, Aid and Influence.* (Washington, DC: Brookings Institution Press, 2008), 59.

22. Jiang, "China's Emerging Strategic Partnerships in Africa," 59.

23. Martyn J. Davies, "Special Economic Zones: China's Developmental Model Comes to Africa" in Robert I. Rotberg, ed., *China Into Africa: Trade, Aid and Influence.* (Washington, DC: Brookings Institution Press, 2008), 142.

24. Davies, "Special Economic Zones," 142.

25. Blanchard, et. al, *Comparing Global Influence,* 128.

26. Blanchard, et. al, *Comparing Global Influence,* 128.

27. Davies, "Special Economic Zones," 142.
28. Blanchard, et. al, *Comparing Global Influence,* 129.
29. Blanchard, et. al, *Comparing Global Influence,* 129.
30. Ministry of External Affairs, Government of India, "India–Angola Relations," July 2013, accessed October 11, 2013, http://www.mea.gov.in/Portal/ForeignRelation/India-Angola_Relations.pdf.
31. "African Union backs India's UNSC bid," *Express News Service,* July 7, 2007, accessed October 11, 2013, http://www.indianexpress.com/news/african-union-backs-indias-unsc-bid/204068/.
32. Embassy of Republic of Angola, Japan website, "Embassy: Message from the Ambassador," (n.d.), accessed October 11, 2013, http://www.angola.or.jp/english/index.php/embassy.
33. Embassy of Republic of Angola, Japan website, "Embassy: Message from the Ambassador," (n.d.), accessed October 11, 2013, http://www.angola.or.jp/english/index.php/embassy.
34. Alex Vines and Indira Campos, "China and India in Angola" in Fantu Cheru and Cyril Obi, eds., *The Rise of China and India in Africa: Challenges, Opportunities and Critical Interventions* (New York, NY: Zed Books, 2010), 195.
35. Ministry of External Affairs, Government of India, "India–Nigeria Relations," January 2012, accessed October 11, 2013, http://www.mea.gov.in/Images/pdf/Nigeria-January-2012.pdf.
36. Ministry of External Affairs, Government of India, "India–Nigeria Relations," January 2012, accessed October 11, 2013, http://www.mea.gov.in/Images/pdf/Nigeria-January-2012.pdf.
37. Ian Taylor, *China's New Role in Africa* (London: Lynne Rienner Publishers, 2009), 47–48.
38. Taylor, *China's New Role in Africa,* 49.
39. Taylor, *China's New Role in Africa,* 49.
40. Obi, "African Oil in the Energy Security Calculations of China and India," 185.
41. Ministry of External Affairs, Government of India, "India–Sudan Relations," January 2013, accessed April 6, 2013, http://www.mea.gov.in/Portal/ForeignRelation/Sudan_Relations__Unclass__January_2013.pdf.
42. Ministry of External Affairs, Government of India, "India–Sudan Relations," January 2013, accessed April 6, 2013, http://www.mea.gov.in/Portal/ForeignRelation/Sudan_Relations__Unclass__January_2013.pdf.
43. Daniel Large and Luke Patey, "Sudan Looks East," in Daniel Large and Luke Patey, eds., *Sudan Looks East: China, India and The Politics of Asian Alternatives* (Rochester, NY: James Currey Ltd., 2011), 7.
44. Ministry of External Affairs, Government of India, "India–Sudan Relations," January 2013, accessed April 6, 2013, http://www.mea.gov.in/Portal/ForeignRelation/Sudan_Relations__Unclass__January_2013.pdf.
45. Ministry of External Affairs, Government of India, "India–Sudan Relations," January 2013, accessed April 6, 2013, http://www.mea.gov.in/Portal/ForeignRelation/Sudan_Relations__Unclass__January_2013.pdf.
46. Ministry of External Affairs, Government of India, "India–South Sudan Relations," May 2012, accessed October 11, 2013, http://www.mea.gov.in/Portal/ForeignRelation/south-sudan-english.pdf.

47. Ministry of External Affairs, Government of India, "India–South Sudan Relations," May 2012, accessed October 11, 2013, http://www.mea.gov.in/Portal/ForeignRelation/south-sudan-english.pdf.

48. Ministry of External Affairs, Government of India, "India–South Sudan Relations," May 2012, accessed October 11, 2013, http://www.mea.gov.in/Portal/ForeignRelation/south-sudan-english.pdf.

49. Ministry of External Affairs, Government of India, "India–South Sudan Relations," May 2012, accessed October 11, 2013, http://www.mea.gov.in/Portal/ForeignRelation/south-sudan-english.pdf.

50. U.S. Energy Information Administration, U.S. Department of Energy, "Country Analysis Briefs: Sudan and South Sudan," March 19, 2012, accessed October 11, 2013, http://www.eia.gov/countries/analysisbriefs/cabs/Sudan/pdf.pdf.

51. Ministry of Foreign Affairs of Japan, "Japan–Sudan Relations (Basic Data)," September 2012, accessed October 12, 2013, http://www.mofa.go.jp/region/africa/sudan/data.html#basic.

52. Murakami Yasuhito, *Japan's Assistance in Peacebuilding in Sudan and Its Challenges* (Washington, DC: Institute of Foreign Policy Analysis, 2011), accessed October 12, 2013, http://www.ifpa.org/pdf/PeaceBuildingApr2011/ Japan Sudan Murakami.pdf.

53. Yasuhito, *Japan's Assistance in Peacebuilding in Sudan and Its Challenges.*

54. Ministry of Foreign Affairs of Japan, "Establishment of Diplomatic Relations with the Republic of South Sudan," July 11, 2011, accessed October 12, 2013, http://www.mofa.go.jp/announce/announce/2011/7/0711_01.html.

55. Jeremy Taylor, "Oil Politics, Asian Suitors, and Alternative Pipelines in South Sudan," *Think Africa Press*, June 26, 2013, accessed October 11, 2013, http://thinkafricapress.com/south-sudan/oil-politics-asian-suitors-alternative-pipelines-south-sudan.

56. Taylor, "Oil Politics, Asian Suitors, and Alternative Pipelines in South Sudan."

57. U.S. Energy Information Administration, U.S. Department of Energy, "Country Analysis Briefs: Sudan and South Sudan," September 5, 2013, accessed October 11, 2013, http://www.eia.gov/countries/analysisbriefs/Sudan/sudan.pdf.

58. U.S. Energy Information Administration, U.S. Department of Energy, "Country Analysis Briefs: Sudan and South Sudan," March 19, 2012, accessed October 11, 2013, http://www.eia.gov/countries/analysisbriefs/cabs/Sudan/pdf.pdf.

59. He Wenping, "The Darfur Issue: A New Test for China's Africa Policy" in Fantu Cheru and Cyril Obi, eds., *The Rise of China and India in Africa: Challenges, Opportunities and Critical Interventions* (New York, NY: Zed Books, 2010), 159.

60. Large and Patey, "Sudan Looks East," 24.

61. Large and Patey, "Sudan Looks East," 24.

62. Large and Patey, "Sudan Looks East," 27.

63. U.S. Energy Information Administration, U.S. Department of Energy, "Country Analysis Briefs: Sudan and South Sudan," March 19, 2012, accessed October 11, 2013, http://www.eia.gov/countries/analysisbriefs/cabs/Sudan/pdf.pdf.

64. U.S. Energy Information Administration, U.S. Department of Energy, "Country Analysis Briefs: Sudan and South Sudan," March 19, 2012, accessed October 11, 2013, http://www.eia.gov/countries/analysisbriefs/cabs/Sudan/pdf.pdf.

65. Oliver Morrissey and Evious Zgovu, *The Impact of China and India on Sub-Saharan African Countries: Opportunities, Challenges and Policies* (London, United Kingdom: Commonwealth Secretariat, 2011), 1, 3.

66. Davies, "Special Economic Zones," 137.

67. Morrissey and Zgovu, *Impact of China and India,* 3.

68. Bates Gill, Chin-hao Huang, and J. Stephen Morrison, *China's Expanding Role in Africa: Implications for the United States* (Washington, DC: CSIS Press, 2007), 8.

69. Gill, Huang and Morrison, *China's Expanding Role in Africa,* 8.

70. Obi, "African Oil in the Energy Security Calculations of China and India," 181.

71. Naidu, "India's African Relations?," 39.

72. Naidu, "India's African Relations?," 40.

73. Naidu, "India's African Relations?," 40–41.

74. Tukumbi Lumumba-Kasongo, *Japan–Africa Relations* (New York, NY: Palgrave Macmillan, 2010), 201.

75. Lumumba-Kasongo, *Japan–Africa Relations,* 202.

76. Lumumba-Kasongo, *Japan–Africa Relations,* 239.

77. Ministry of External Affairs, Government of India, "India–Angola Relations," July 2013, accessed October 11, 2013, http://www.mea.gov.in/Portal/ForeignRelation/India-Angola_Relations.pdf.

78. Ministry of External Affairs, Government of India, "India–Angola Relations," July 2013, accessed October 11, 2013, http://www.mea.gov.in/Portal/ForeignRelation/India-Angola_Relations.pdf.

79. U.S. Energy Information Administration, U.S. Department of Energy, "Country Analysis Briefs: Angola," January 8, 2013, accessed October 11, 2013, http://www.eia.gov/countries/analysisbriefs/Angola/angola.pdf.

80. Ministry of External Affairs, Government of India, "India–Angola Relations," July 2013, accessed October 11, 2013, http://www.mea.gov.in/Portal/ForeignRelation/India-Angola_Relations.pdf.

81. Ministry of External Affairs, Government of India, "India–Angola Relations," July 2013, accessed October 11, 2013, http://www.mea.gov.in/Portal/ForeignRelation/India-Angola_Relations.pdf.

82. Ministry of External Affairs, Government of India, "India–Angola Relations," July 2013, accessed October 11, 2013, http://www.mea.gov.in/Portal/ForeignRelation/India-Angola_Relations.pdf.

83. Ministry of External Affairs, Government of India, "India–Angola Relations," July 2013, accessed October 11, 2013, http://www.mea.gov.in/Portal/ForeignRelation/India-Angola_Relations.pdf.

84. Ana Cristina Alves, *The Oil Factor in Sino-Angolan Relations at the Start of the 21st Century* (Johannesburg, South Africa: South African Institute of International Affairs), February 2010, accessed October 12, 2013, http://www.voltairenet.org/IMG/pdf/Sino-Angolan_Relations.pdf.

85. Ministry of Foreign Affairs of Japan, "Japan–Angola Relations (Basic Data)," May 2009, accessed October 13, 2013, http://www.mofa.go.jp/region/africa/angola/data.html.

86. Ministry of Foreign Affairs of Japan, "Japan–Angola Relations (Basic Data)," May 2009, accessed October 13, 2013, http://www.mofa.go.jp/region/africa/angola/data.html.

87. Ministry of Foreign Affairs of Japan, "Japan–Angola Relations (Basic Data)," May 2009, accessed October 13, 2013, http://www.mofa.go.jp/region/africa/angola/data.html.

88. Obi, "African Oil in the Energy Security Calculations of China and India," 184.

89. "China pledges $20bn in credit for Africa at summit," *BBC News China*, July 19, 2012, accessed October 13, 2012, http://www.bbc.co.uk/news/world-asia-china-18897451.

90. Ron Sandrey, *The Impact of China–Africa Trade Relations: The Case of Angola* (paper prepared for the African Economic Research Consortium, Nairobi, Kenya), November 2009, accessed October 13, 2013, http://dspace.cigilibrary.org/jspui/bitstream /123456789/32375/1/Angola-TradeStudy.pdf?1.

91. Vines and Campos, "China and India in Angola," 200.

92. Vines and Campos, "China and India in Angola," 200.

93. Vines and Campos, "China and India in Angola," 205.

94. Lucy Corkin, "China and Angola: Strategic partnership or marriage of convenience?" (Bergen, Norway: Chr. Michelsen Institute), January 2011, Vol. 1, No. 1, accessed October 13, 2013, http://www.cmi.no/publications/file/3938-china-and-angola-strategic-partnership-or-marriage.pdf.

95. Corkin, "China and Angola?"

96. Alves, *The Oil Factor in Sino-Angolan Relations*."

97. Alves, *The Oil Factor in Sino-Angolan Relations*."

98. Ministry of External Affairs, Government of India, "India–Nigeria Relations," July 2013, accessed October 11, 2013, http://www.mea.gov.in/Portal/ForeignRelation/India-Nigeria_Relations.pdf.

99. Ministry of External Affairs, Government of India, "India–Nigeria Relations," July 2013, accessed October 11, 2013, http://www.mea.gov.in/Portal/ForeignRelation/India-Nigeria_Relations.pdf.

100. Ministry of External Affairs, Government of India, "India–Nigeria Relations," July 2013, accessed October 11, 2013, http://www.mea.gov.in/Portal/ForeignRelation/India-Nigeria_Relations.pdf.

101. Ministry of External Affairs, Government of India, "India–Nigeria Relations," January 2012, accessed October 13, 2013, http://www.mea.gov.in/Images/pdf/Nigeria-January-2012.pdf.

102. Ministry of External Affairs, Government of India, "India–Nigeria Relations," January 2012, accessed October 13, 2013, http://www.mea.gov.in/Images/pdf/Nigeria-January-2012.pdf.

103. Ministry of Foreign Affairs of Japan, "Japan–Nigeria Relations (Basic Data)," accessed October 13, 2013, http://www.mofa.go.jp/region/africa/nigeria/data.html.

104. Ministry of Foreign Affairs of Japan, "Japan–Nigeria Relations (Basic Data)," accessed October 13, 2013, http://www.mofa.go.jp/region/africa/nigeria/data.html.

105. Dapo Fafowora, "Nigeria's growing relations with China," *The Nation*, July 18, 2013, accessed October 13, 2013, http://thenationonlineng.net/new/nigerias-growing-economic-relations-with-china/.

106. Djeri-wake Nabine, *The Impact of Chinese Investment and Trade on Nigeria Economic Growth* (African Trade Policy Centre, Economic Commission for Africa),

2009, accessed October 13, 2013, http://www.uneca.org/sites/default/files/publications/ 77.pdf.

107. Dapo Fafowora, "Nigeria's growing relations with China," *The Nation*, July 18, 2013, accessed October 13, 2013, http://thenationonlineng.net/new/nigerias-growing-economic-relations-with-china/.

108. Dapo Fafowora, "Nigeria's growing relations with China," *The Nation*, July 18, 2013, accessed October 13, 2013, http://thenationonlineng.net/new/nigerias-growing-economic-relations-with-china/.

109. Ministry of External Affairs, Government of India, "India–Sudan Relations," August 2013, accessed October 11, 2013, http://www.mea.gov.in/Portal/ForeignRelation/INDIA-SUDAN_RELATIONS.pdf.

110. Ministry of External Affairs, Government of India, "India–Sudan Relations," August 2013, accessed October 11, 2013, http://www.mea.gov.in/Portal/ForeignRelation/INDIA-SUDAN_RELATIONS.pdf.

111. U.S. Energy Information Administration, U.S. Department of Energy, "Country Analysis Briefs: Sudan and South Sudan," March 19, 2012, accessed October 11, 2013, http://www.eia.gov/countries/analysisbriefs/cabs/Sudan/pdf.pdf.

112. U.S. Energy Information Administration, U.S. Department of Energy, "Country Analysis Briefs: Sudan and South Sudan," March 19, 2012, accessed October 11, 2013, http://www.eia.gov/countries/analysisbriefs/cabs/Sudan/pdf.pdf.

113. Ministry of External Affairs, Government of India, "India–South Sudan Relations," May 2012, accessed October 11, 2013, http://www.mea.gov.in/Portal/ForeignRelation/south-sudan-english.pdf.

114. Ministry of Foreign Affairs of Japan, "Japan–Iran Relations," September 2012, accessed October 12, 2013, http://www.mofa.go.jp/region/africa/sudan/data.html#basic.

115. Ministry of Foreign Affairs of Japan, "Japan–Iran Relations," September 2012, accessed October 12, 2013, http://www.mofa.go.jp/region/africa/sudan/data.html#basic.

116. U.S. Energy Information Administration, U.S. Department of Energy, "Country Analysis Briefs: Sudan and South Sudan," March 19, 2012, accessed October 11, 2013, http://www.eia.gov/countries/analysisbriefs/cabs/Sudan/pdf.pdf.

117. U.S. Energy Information Administration, U.S. Department of Energy, "Country Analysis Briefs: Sudan and South Sudan," March 19, 2012, accessed October 11, 2013, http://www.eia.gov/countries/analysisbriefs/cabs/Sudan/pdf.pdf.

118. U.S. Energy Information Administration, U.S. Department of Energy, "Country Analysis Briefs: Sudan and South Sudan," March 19, 2012, accessed October 11, 2013, http://www.eia.gov/countries/analysisbriefs/cabs/Sudan/pdf.pdf.

119. "China pledges $20bn in credit for Africa at summit," *BBC News China*, July 19, 2012, accessed October 13, 2012, http://www.bbc.co.uk/news/world-asia-china-18897451.

120. Leben Nelson Moro, "China, Sudan and South Sudan Relations," *Global Review*, Winter 2012, accessed October 13, 2013, http://www.saferworld.org.uk/downloads/pubdocs/Leben-Moro.pdf.

121. Wenping, "The Darfur Issue," 159.

122. Wenping, "The Darfur Issue," 159.

123. Wenping, "The Darfur Issue," 159.

124. Embassy of the People's Republic of China in South Sudan website, "Sino-South Sudan Economic & Trade Cooperation," November 23, 2012, accessed October 13, 2013, http://ss.chineseembassy.org/eng/sbgx/jmwl/.

125. Embassy of the People's Republic of China in South Sudan website, "Sino-South Sudan Economic & Trade Cooperation," November 23, 2012, accessed October 13, 2013, http://ss.chineseembassy.org/eng/sbgx/jmwl/.

126. Embassy of the People's Republic of China in South Sudan website, "Sino-South Sudan Economic & Trade Cooperation," November 23, 2012, accessed October 13, 2013, http://ss.chineseembassy.org/eng/sbgx/jmwl/.

127. "S. Sudan says Japan to construct alternative oil pipeline," *Sudan Tribune*, June 1, 2013, accessed October 13, 2013, http://www.sudantribune.com/spip.php?article46786.

128. David H. Shinn, "Military and Security Relations: China, Africa, and the Rest of the World" in Robert I. Rotberg, ed., *China Into Africa: Trade, Aid, and Influence*. (Washington, DC: Brookings Institution Press, 2008), 154.

129. Shinn, "Military and Security Relations," 177.

130. Shinn, "Military and Security Relations," 180.

131. Shinn, "Military and Security Relations," 183.

132. SIPRI Arms Transfers Database, <http://www.sipri.org/databases/armstransfers>.

133. SIPRI Arms Transfers Database, <http://www.sipri.org/databases/armstransfers>.

134. Bruno De Paiva, "Japan's National Involvement in the Indian Ocean Region," *Future Directions International*, July 29, 2011.

135. De Paiva, "Japan's National Involvement in the Indian Ocean Region."

136. Shinn, "Military and Security Relations," 165.

137. Shinn, "Military and Security Relations," 165.

138. Tanvi Madan, *The Brookings Foreign Policy Studies Energy Security Series: India* (Washington, DC: Brookings Institution Press), November 2007, accessed October 11, 2013, http://www.brookings.edu/~/media/research/files/reports/2006/11/india/2006india.pdf.

139. SIPRI Arms Transfers Database, <http://www.sipri.org/databases/armstransfers>.

140. SIPRI Arms Transfers Database, <http://www.sipri.org/databases/armstransfers>.

7

Europe and Eurasia

In terms of the states that produce crude oil, this region is comprised of Azerbaijan, Norway, the United Kingdom and Russia. Of these, only Azerbaijan and Russia export oil to China, India, and Japan. Because China and Japan have energy stakes in Azerbaijan and not India, only Russia will be discussed. See appendices B, C, and D.

Many observers do not see Europe and Eurasia as an area of either Sino-Indian or Sino-Japanese competition for oil energy resources because the region holds approximately 1 percent of world's proven oil reserves. Further, both India and Japan have shared values with the European Union (EU) such as democracy and the rule of law.[1]

India and Japan also have robust trade relationships with the European Union. The European Union is India's largest trading partner.[2] Indo-European Union bilateral trade was $102.5 billion (€78.5 billion) in 2012.[3] The European Union is also one of the largest sources of foreign direct investment to India.[4]

The European Union is also Japan's third-largest trading partner.[5] Japanese–European Union bilateral trade was $156 billion (€119.3 billion) in 2012.[6] Japan is also a net investor of foreign direct investment in the European Union. In 2011, Japanese foreign direct investment was $15.6 billion (€12.068 billion).[7]

India and Japan have military relations with the European Union. India promotes defense dialogue, cooperation and exchanges with the Czech Republic, France, Germany, Norway, Slovakia, Turkey, and the United Kingdom.[8] Moreover, Japan promotes defense dialogue, cooperation and exchanges with Canada, France, Germany, Italy, Turkey, and the United

Kingdom.[9] Both India and Japan have had historically close military relations with the United Kingdom, which include reciprocal defense attachés.

In terms of Russian annual crude oil exports, approximately 1 percent is exported to India and approximately 5 percent is exported to Japan according to FACTS Global Energy. Comparatively, China imports approximately 8 percent of its crude oil from Europe and Eurasia according to FACTS Global Energy. Thus, the region ranks last in the world as an oil producer to China.

The PRINCE model predicts that China will continue to import oil from Europe and Eurasia based on a probability of 95 percent or greater.[10]

EUROPE: POLITICAL, ECONOMIC, AND MILITARY IMPLICATIONS FOR INDIA AND JAPAN

Sino-Western European diplomatic relations have improved steadily since the end of the Cold War as evidenced by high-level political exchanges and regular meetings including political summits involving the national leaders of China and the European Union. Further, Western Europe and China have engaged in economic, energy, environmental, scientific and technological and educational and legal cooperation to strengthen bilateral ties. However, China and Western Europe have yet to forge a strategic partnership. Persistent concerns over China's authoritarian nature, legal system and human rights record impede further development and deepening of bilateral diplomatic relations despite some considerable progress over the last twenty years, especially since June 1989 when China cracked down on political activists at Tiananmen Square, which resulted in an arms embargo on China and tensions in bilateral relations. In contrast, India and Japan have shared values with Western Europe.

Western Europe's close ties with Japan's ally, the United States, also factor in the development of Sino-Western European diplomatic relations. Western Europe is aligned with the United States having a long tradition of strategic cooperation that extends beyond the political realm and into the economic and military ones and shared commitments to a number of basic values that transcend Sino-Western European relations despite political tensions stemming from America's past unilateralist approach to international relations such as the invasion of Iraq and the treatment of unlawful combatants.[11] Therefore, China's energy relations with Western Europe, which mainly consists of importing crude oil from Norway and the United Kingdom, is unlikely to result in any diplomatic competition for oil energy resources between China and India or China and Japan.

Trade and economic ties appear to be the foundation of the Sino-Western European relationship even though China has sought to foster "greater European political and strategic independence from the United States."[12] Western Europe (under the auspices of the European Union) has been China's largest trading partner and the European Union has been China's second-largest trading partner after the United States since 2004.[13] In contrast, the European Union is India's largest trading partner and Japan's third-largest trading partner.

Although trade and economic relations have deepened between China and Western Europe (under the auspices of the European Union), it has not been without friction.[14] Since late 2006, trade and economic problems have played a more prominent role in Sino-Western European economic relations across a wide range of issues: European Union trade deficit, unfair trade practices, intellectual property rights violations, currency valuation and the export of dangerous and tainted goods by China.[15] Further, Europe also refused to extend "Market Economy Status" (MES) to China because of unfair trade practices even though China joined the World Trade Organization in 2001.[16] Apart from persistent political issues in the relationship such as China's governability and human rights record, these issues have prevented China and Western Europe from implementing a strategic economic relationship.[17]

Western Europe's strategic relationship with the United States also factors in Sino-Western European economic relations. Despite past differences in foreign policy approaches, Western Europe and the United States have deeply integrated economies (not least in defense sector) and shared commitments to a number of basic values.[18] China's energy relations with Western Europe, which mainly consists of importing crude oil from Norway and the United Kingdom, is unlikely to result in economic competition for oil energy resources between China and India or China and Japan. India and Japan import very little if any crude oil from Norway or the United Kingdom.

Similarly to political and economic relations, China's oil energy trading in Western Europe (which is marginal at best) is unlikely to lead to a military competition for strategic energy resources because of the strong military relationships that exist between the United States and the European Union. The United States has had a far longer and deeper military relationship with Western Europe than China especially since 1949 when the North Atlantic Treaty Organization (NATO) was formed for the collective defense of its members in the event of an attack by an external party. Participation as a member in NATO means participation in hard military activities such as coalition warfare as opposed to soft military activities such as military attaché exchanges, educational exchanges, port visits, observing military exercises, or even UN peacekeeping. Note that

China historically has never subscribed to formal military alliances even with its erstwhile communist ally Russia. China has provided military aid and support to regimes that are inimical to U.S. interests such as Burma, Iran, and the Sudan. However, these actions have not translated into a formal military alliance. Overall, Western Europe is formally aligned to the United States militarily.

China's oil energy trading in Europe and Eurasia is unlikely to lead to a military competition for strategic energy resources because the region is not vital to China's energy security compared to the Near East and Africa. In this respect, China imports only less than 0.2 percent of its oil from Western Europe compared to Africa and the Near East, which is respectively 30 percent and 50 percent. Thus, Europe and Eurasia is strategically insignificant in terms of energy security compared to the other oil producing regions of the world.

RUSSIA

The PRINCE model predicts that an oil pipeline will be built from Siberia into Northern China by 94 percent and that China and Russia will jointly explore for oil in the Arctic Circle by 98 percent in the near future.[19]

China–Russia–India–Japan Diplomatic/Political Relations

Indo-Russian diplomatic relations are strategic. Since the signing of the "Declaration on the India–Russian Strategic Partnership" in October 2000, bilateral ties have acquired a qualitatively new character with enhanced levels of cooperation in almost all areas: political, security, trade and economy, defense, science and technology, and culture.[20] India and Russia have also conducted thirteen annual summits to date to solidify their strategic relationship.[21] The annual summits are considered the highest institutionalized dialogue mechanism between the two countries.[22]

Japanese–Russian diplomatic relations are ongoing and are characterized by the mutual adherence to democratic values, the absence of ideological and military confrontation, and a mutual interest in international cooperation.[23] However, Russia is a troubled democracy in reality.

Although more than sixty years have passed since the end of the Second World War, Japan and Russia have yet to conclude a peace treaty. The reason is that an unresolved territorial issue exists between them concerning the Northern Territories (the islands of Etorofu, Kunashiri, Shikotan and Habomai).[24] Russia occupies and controls them and Japan considers them inherent territories handed down from generation to generation without ever being territories of another country.[25]

Sino-Russian diplomatic relations are robust but have had a long and often troubled history.[26] Since 1990 their relationship could be considered successful if our observations are confined to top-level political relations and recognize how China and Russia eventually resolved their political disputes, settled border issues, minimized the risk of armed conflict, and eventually formed a strategic partnership relationship in 1996.[27]

Energy has the potential to further strengthen long-term diplomatic ties between China and Russia. In this respect, Transneft completed a forty mile oil pipeline spur from Skovorodino, Siberia to the Amur River on Russia–China border followed by the China National Petroleum Corporation (CNPC) completing a 616 mile section from Russia–China border to Daqing, China in October 2010. Operations began on January 1, 2011.

Transneft also opened the East Siberian Pacific Ocean (ESPO) oil pipeline on December 25, 2012. The first leg of the oil pipeline runs 1,713 miles from Taishet to Skovorodino (in the Amur Region of Russia's Far East) and the second leg of the oil pipeline continues for 1,304 miles from Skovorodino to the Pacific Port of Kozmino on Russia's west coast. This oil pipeline is designed to pump annually 30 million metric tons of Siberian oil to China, Japan, and South Korea.[28]

In terms of future strategic energy cooperation, China's largest oil company, CNPC, and Rosneft, one of three major state-run Russian companies, may jointly explore for oil in the Kara Sea in the Arctic Circle in the near future. Oil exploration in the Kara Sea would be a major energy coup for both countries, but even more so for Russia because it could elevate its position from the world's second-largest oil exporter to the first. Thus, while potentially promising, this event is highly speculative because no agreements have been signed to date. However, with an event occurrence probability of 98 percent, there is a very good chance that this event may occur in the near future. This would give China greater oil energy security given the geographic location of the event and its geographic inaccessibility to other major oil regions of the world.

Sino-Russian oil energy relations are unlikely to have any serious political consequences for either India or Japan. China is more likely to be affected in terms of energy security because it is more oil energy dependent on Russia than either India or Japan. In this respect, 8 percent of China's total crude oil and petroleum imports are from Russia compared to India's 1 percent and Japan's 5 percent according to FACTS Global Energy. Despite China's energy insecurity, it is unlikely Japanese–Russian relations will improve in the short-term because of unresolved territorial sovereignty issues regarding the Northern Territories.

China–Russia–India–Japan Economic Relations

Indo-Russian economic relations are robust. Bilateral trade has grown from $7.5 billion in 2009 to $11.04 billion in 2012.[29] Between January and July 2013, bilateral trade amounted to $3.9 billion.[30] The two-way investment between the two countries stands at approximately $7.8 billion.[31]

In terms of bilateral trade, India exports to Russia pharmaceuticals; tea, coffee and spices; apparel and clothing; edible preparations; and engineering goods.[32] Russia exports to India iron and steel; fertilizers; non-ferrous metals; paper products; coal, coke and briquettes; cereals; and rubber.[33]

India and Russia have also established a number of institutional mechanisms to supervise bilateral economic cooperation and promote direct bilateral business to business contacts. The main institutional mechanism supervising economic cooperation is the India–Russia Inter-Governmental Commission on Trade, Economic, Scientific, Technological and Cultural Cooperation (IRIGC-TEC).[34] The IRIGC-TEC integrates six (6) working groups on economic and trade cooperation, modernization and industrial cooperation, energy, tourism and culture, science and technology, and IT.[35] Moreover, the India–Russia Forum on Trade and Investment and India–Russia CEOs' Council are the two primary mechanisms to promote direct bilateral business-to-business contacts between India and Russia.[36]

In terms of energy stakes, India's OVL acquired Imperial Energy in August 2008, which through its subsidiaries owns the licenses of the blocks/fields in Russia. OVL also acquired a 20 percent stake in the Sakhalin 1 project in 2008. The Sakhalin 1 project is a consortium to explore and produce oil and gas at three fields: Chayvo, Odoptu and Arkutun-Dagi (offshore Sakhalin). See appendix C. Despite these energy stakes, India is only a minor importer of Russian crude oil at 1 percent annually.

Japanese–Russian economic relations are robust. Bilateral trade increased almost 13 percent in 2012 and reached $33.5 billion.[37] While significant, Japan is Russia's fourteenth-largest trading partner while Russia is Japan's ninth-largest trading partner.[38]

In terms of direct investment, Japan invested $10.7 billion into Russia in 2012.[39] Japan is considered one of the top ten leaders of foreign investment in Russia.[40]

In terms of bilateral trade, Japan exports to Russia automobiles; trucks; auto parts; construction equipment; printing equipment; radio and television equipment; medical and power-generation equipment; and lift-and-carry machinery.[41] Russia exports to Japan crude oil, liquefied natural gas;

oil products; unprocessed aluminum; black coal; unprocessed palladium and platinum; lumber; ferrous alloy; and marine products.[42]

Japan and Russia have also established a number of institutional mechanisms to supervise bilateral economic cooperation and promote direct bilateral business to business contacts. The main institutional mechanism to promote bilateral economic relations is the Russia–Japan Intergovernmental Committee on Trade and Economic Issues.[43] Moreover, the Russia–Japan and Japan–Russia Committees on Economic Cooperation promote bilateral ties at the business level.[44]

In terms of energy stakes, Japan's Sakhalin Oil and Gas Development Co., Ltd. (SODECO) acquired a 30 percent interest in the Sakhalin 1 project in 2007. See appendix D. Japan is also a modest importer of Russian crude oil at 5 percent annually.

Sino-Russian economic relations are robust. Bilateral trade was $88.16 billion in 2012.[45] Bilateral trade is expected to reach $100 billion in 2014.[46] China is also Russia's largest trading partner.[47]

Foreign direct investment is relatively small between China and Russia. In 2011, Chinese FDI in Russia was $3.76 billion while Russian FDI in China was $123.1 million.[48] FDI between BRIC countries like China and Russia appears to be limited according to the United Nations Conference on Trade and Development (UNCTAD).[49]

In terms of bilateral trade, China exports to Russia shoes and garments, and machinery and equipment which include automobiles, computers, telephones and electronics.[50] Russia mainly exports to China crude oil and raw materials, food, chemical products, and base metals.[51] Russia also exports arms to China.[52]

In terms of energy stakes, China's Sinopec purchased 97 percent of Udmurtneft for $1.7 billion in 2006; CNPC purchased a 49 percent stake in Vostok Energy Ltd. in October 2006; and CIC purchased 45 percent of Nobel Oil Group for $300 million in October 2009. See appendix B. China is also a modest importer of Russian crude oil at 8 percent annually.

Sino-Russian energy relations will not adversely affect India or Japan economically either regionally or globally because India or Japan do not depend on Russia for the majority of its crude petroleum. In this respect, Russia is only a minor to modest exporter of crude petroleum to India and Japan compared to China. Further, Russia seeks to diversify its hydrocarbon energy market by making its crude oil and petroleum available not only to China but also to India and Japan.[53] This is evidenced by the operation of the East Siberian Pacific Ocean (ESPO) oil pipeline from Taishet to the Pacific Port of Kozmino on Russia's east coast in December 2012. Therefore, it is highly unlikely that India and Japan will be impacted economically.

China–Russia–India–Japan Military Relations

Indo-Russian military relations are good. According to the Indian Ministry of Defence, India and Russia have a long standing defense relationship based on mutual trust and understanding.[54] In terms of defense cooperation, Russia continues to remain an important supplier of military equipment to India.[55] Examples of Russian arms exported to India are warships, fighter aircraft, anti-ship missiles, and tanks according to SIPRI. It is the only country that India has an institutionalized annual defense cooperation mechanism at the defense minister level.[56] India and Russia have also conducted joint military exercises since 2003.[57]

Indo-Russian defense technical cooperation has evolved from a simple buyer-seller framework to one involving joint research and development, joint production and marketing of advanced defense technologies and systems.[58] Examples of flagship cooperation are the BrahMos Missile System, joint development of the fifth generation fighter aircraft and the multi-transport aircraft, as well as the licensed production in India of SU-30 aircraft and T-90 tanks.[59]

Japanese–Russian military relations are continuously developing with the goal of establishing mutual trust and cooperation. According to the Japanese Ministry of Defense, Japan and Russia promote military exchanges at various levels in accordance with the 1999 memorandum on Japan–Russia Defense Exchanges.[60] Security consultations between foreign and defense authorities, and bilateral military talks, as well as annual meetings based on the Japan–Russia Agreement on Prevention of Incidents On and Over the High Seas and joint search and rescue exercises are held continuously.[61]

In recent years, Japanese–Russian defense exchanges have not been active partly because of the reorganization of the Russian Army.[62] However, Japanese Maritime Self Defense Force (MSDF) and the Russian Navy held the twelfth Japan–Russia Joint Search and Rescue Exercise and Russian naval vessels visited Japan in September 2011.[63]

Sino-Russian military relations are generally good but are characterized by some mistrust. While China and Russia do not have a formal military alliance, they cooperate on defense matters. In this respect, there are three dimensions of bilateral military cooperation: military-political cooperation in the form of high-level meetings between military chiefs of staff and defense ministers; military training cooperation via exchanges of personnel for training, military education, and joint military exchanges; and military-technical cooperation through the transfer of arms, technology, and know-how for the production of military equipment.[64]

In terms of military training cooperation, China and Russia have conducted joint military exercises to strengthen joint operational capabilities, exchange experience, facilitate cooperation in the fight against the "three forces" of terrorism, separatism and extremism, and enhance mutual combat readiness against emerging threats.[65] The joint military exercises are peace mission exercises and may include members of the Shanghai Cooperation Organization. Peace mission exercises were conducted in 2005, 2007, 2009, 2010, 2012, and 2013.

In terms of military cooperation, the transfer of major conventional weapons, components, and technologies from Russia to China has been a cornerstone of the Sino-Russian relationship. This was very apparent between 1991 and 2010 when 90 percent of China's imported major conventional weapons were supplied by Russia.[66] During this period China imported from Russia Su-27/Su-30 combat aircraft, transport aircraft, Mi-17 military transport helicopters, Tor-M1 mobile air defense systems, S-300PMU1/2 air defense systems, Type 636E and Type 877E submarines, Sovremenny destroyers and a wide range of missiles. In addition, China secured agreement for the licensed production of Su-27 combat aircraft, Mi-17 helicopters and anti-tank and anti-ship missiles according to SIPRI.[67] However, since 2007 there has been a notable decline in Russian arms deliveries to China.[68]

A strong Sino-Russian oil energy partnership will not adversely affect India or Japan militarily either regionally or globally or lead to a military competition for strategic energy resources because India or Japan do not depend on Russia for the majority of its crude petroleum. Russia also seeks to make its hydrocarbons available to India and Japan in addition to China.

NOTES

1. Ministry of Foreign Affairs of Japan, "Japan–EU Relations," January 2013, accessed October 13, 2013, http://www.mofa.go.jp/region/europe/eu/pdfs/j-eu-relations.pdf.

2. Ministry of External Affairs, Government of India, "India–EU Relations," July 2013, accessed October 13, 2013, http://mea.gov.in/Portal/ForeignRelation/India-EU_Relations.pdf.

3. Ministry of External Affairs, Government of India, "India–EU Relations," July 2013, accessed October 13, 2013, http://mea.gov.in/Portal/ForeignRelation/India-EU_Relations.pdf.

4. Ministry of External Affairs, Government of India, "India–EU Relations," July 2013, accessed October 13, 2013, http://mea.gov.in/Portal/ForeignRelation/India-EU_Relations.pdf.

5. European Commission, "A Free Trade Agreement between the EU and Japan," memo 13-572, June 17, 2013, accessed October 13, 2013, http://europa.eu/rapid/press-release_MEMO-13-572_en.htm.

6. Eurostat, "EU27 trade in goods deficit with Japan restricted to 8bn euro in 2012," news release, March 22, 2013, accessed October 13, 2013, http://epp.eurostat.ec.europa.eu/cache/ITY_PUBLIC/6-22032013-AP/EN/6-22032013-AP-EN.PDF.

7. Eurostat, "EU27 trade in goods deficit with Japan restricted to 8bn euro in 2012," news release, March 22, 2013, accessed October 13, 2013, http://epp.eurostat.ec.europa.eu/cache/ITY_PUBLIC/6-22032013-AP/EN/6-22032013-AP-EN.PDF.

8. Ministry of Defence, Government of India, *Annual Report 2011–2012*, 2012, 196–197.

9. Japanese Ministry of Defense, *Defense of Japan 2012*, 2012, 279.

10. Prince and Panda website, "PRINCE_SCORES.pdf," accessed October 17, 2013, http://www.princeandpanda.com.

11. Andrew Small, "The US Factor", in Stanley Crossick and Etienne Reuter, eds., *China–EU: A Common Future* (Singapore: World Scientific Publishing, 2007), 74.

12. Robert G. Sutter, *Chinese Foreign Relations: Power and Policy Since the Cold War* (Lanham, MD: Rowman & Littlefield Publishers, 2008), 344–345.

13. Sutter, *Chinese Foreign Relations*, 344–345.

14. Bates Gill and Melissa Murphy, *China–Europe Relations: Implications and Policy Reponses for the United States* (Washington, DC: CSIS Press, May 2008), 10.

15. Gill and Murphy, *China–Europe Relations*, 10–11.

16. David Shambaugh, "China eyes Europe in the world: real convergence or cognitive dissonance?" in David Shambaugh, Eberhard Sandschneider, and Zhou Hong, eds., *China–Europe Relations: Perceptions, Policies and Prospects* (New York, NY: Routledge, 2008), 136.

17. Fraser Cameron and Yongnian Zheng, "Key Elements of a Strategic Partnership," in Stanley Crossick and Etienne Reuter, eds., *China–EU: A Common Future*. (Singapore: World Scientific Publishing, 2007), 6.

18. Small, "The US Factor," 74.

19. Prince and Panda website, "PRINCE_SCORES.pdf," accessed October 17, 2013, http://www.princeandpanda.com.

20. Ministry of External Affairs, Government of India, "India–Russia Relations," July 2013, accessed October 13, 2013, http://www.mea.gov.in/Portal/ForeignRelation/India-Russia_Relations.pdf.

21. Ministry of External Affairs, Government of India, "India–Russia Relations," July 2013, accessed October 13, 2013, http://www.mea.gov.in/Portal/ForeignRelation/India-Russia_Relations.pdf.

22. Ministry of External Affairs, Government of India, "India–Russia Relations," July 2013, accessed October 13, 2013, http://www.mea.gov.in/Portal/ForeignRelation/India-Russia_Relations.pdf.

23. The Embassy of the Russian Federation to Japan website, "Russian–Japanese Relations," accessed October 13, 2013, http://www.russia-emb.jp/english/embassy /relations.html.

24. Ministry of Foreign Affairs of Japan, "Japan's Northern Territories: For a Relationship of Genuine Trust," accessed October 13, 2013, http://www.mofa.go.jp/region/europe/russia/territory/pamphlet.pdf.

25. Ministry of Foreign Affairs of Japan, "Japan's Northern Territories: For a Relationship of Genuine Trust," accessed October 13, 2013, http://www.mofa.go.jp/region/europe/russia/territory/pamphlet.pdf.

26. Sutter, *Chinese Foreign Relations*, 327.

27. Matti Nojonen, "Introduction: Adjusting to the Great Power Transition," in Arkady Moshes and Matti Nojonen, eds., *Russia–China Relations: Current State, Alternative Futures, and Implications for the West*. (Helsinki, Finland: The Finnish Institute of Foreign Affairs, 2011), 14.

28. Mikhail Fomichev, "Transneft to Boost Pacific Oil Pipeline Capacity to 67M tons," *Ria Novosti*, July 31, 2013, accessed October 13, 2013, http://en.rian.ru/business/ 20130731/182511026.html.

29. Ministry of External Affairs, Government of India, "India–Russia Relations," July 2013, accessed October 13, 2013, http://www.mea.gov.in/Portal/ForeignRelation/India-Russia_Relations.pdf.

30. Ministry of External Affairs, Government of India, "India–Russia Relations," July 2013, accessed October 13, 2013, http://www.mea.gov.in/Portal/ForeignRelation/India-Russia_Relations.pdf.

31. Ministry of External Affairs, Government of India, "India–Russia Relations," July 2013, accessed October 13, 2013, http://www.mea.gov.in/Portal/ForeignRelation/India-Russia_Relations.pdf.

32. Embassy of India to Russia website, "Overview of India–Russia Economic Cooperation," accessed October 13, 2013, http://indianembassy.ru/index.php?option=com_content&view=article&id=705&Itemid=705&lang=en.

33. Embassy of India to Russia website, "Overview of India–Russia Economic Cooperation," accessed October 13, 2013, http://indianembassy.ru/index.php?option=com_content&view=article&id=705&Itemid=705&lang=en.

34. Ministry of External Affairs, Government of India, "India–Russia Relations," July 2013, accessed October 13, 2013, http://www.mea.gov.in/Portal/ForeignRelation/India-Russia_Relations.pdf.

35. Ministry of External Affairs, Government of India, "India–Russia Relations," July 2013, accessed October 13, 2013, http://www.mea.gov.in/Portal/ForeignRelation/India-Russia_Relations.pdf.

36. Ministry of External Affairs, Government of India, "India–Russia Relations," July 2013, accessed October 13, 2013, http://www.mea.gov.in/Portal/ForeignRelation/India-Russia_Relations.pdf.

37. The Embassy of the Russian Federation to Japan website, "Economic Relations between Russia and Japan," accessed October 13, 2013, http://www.russia-emb.jp/english/embassy/economic.html.

38. The Embassy of the Russian Federation to Japan website, "Economic Relations between Russia and Japan," accessed October 13, 2013, http://www.russia-emb.jp/english/embassy/economic.html.

39. The Embassy of the Russian Federation to Japan website, "Economic Relations between Russia and Japan," accessed October 13, 2013, http://www.russia-emb.jp/english/embassy/economic.html.

40. The Embassy of the Russian Federation to Japan website, "Economic Relations between Russia and Japan," accessed October 13, 2013, http://www.russia-emb.jp/english/embassy/economic.html.

41. The Embassy of the Russian Federation to Japan website, "Economic Relations between Russia and Japan," accessed October 13, 2013, http://www.russia-emb.jp/english/embassy/economic.html.

42. The Embassy of the Russian Federation to Japan website, "Economic Relations between Russia and Japan," accessed October 13, 2013, http://www.russia-emb.jp/english/embassy/economic.html.

43. The Embassy of the Russian Federation to Japan website, "Economic Relations between Russia and Japan," accessed October 13, 2013, http://www.russia-emb.jp/english/embassy/economic.html.

44. The Embassy of the Russian Federation to Japan website, "Economic Relations between Russia and Japan," accessed October 13, 2013, http://www.russia-emb.jp/english/embassy/economic.html.

45. "China–Russia trade volume to hit 100 bln USD in 2014," *English News*, June 14, 2013, accessed October 13, 2013, http://news.xinhuanet.com/english/china/2013-06/14/c_132456315.htm.

46. "China–Russia trade volume to hit 100 bln USD in 2014," *English News*, June 14, 2013, accessed October 13, 2013, http://news.xinhuanet.com/english/china/2013-06/14/c_132456315.htm.

47. "China–Russia trade volume to hit 100 bln USD in 2014," *English News*, June 14, 2013, accessed October 13, 2013, http://news.xinhuanet.com/english/china/2013-06/14/c_132456315.htm.

48. United Nations Conference on Trade and Investment, "The Rise of the BRICS FDI and Africa," March 25, 2013, accessed October 13, 2013, http://unctad.org/en/PublicationsLibrary/webdiaeia2013d6_en.pdf.

49. United Nations Conference on Trade and Investment, "The Rise of the BRICS FDI and Africa," March 25, 2013, accessed October 13, 2013, http://unctad.org/en/PublicationsLibrary/webdiaeia2013d6_en.pdf.

50. Jouko Rautava, "Russia's Economic Policy and Russia–China Economic Relations", in Arkady Moshes and Matti Nojonen, eds., *Russia–China Relations: Current State, Alternative Futures, and Implications for the West*. (Helsinki, Finland: The Finnish Institute of Foreign Affairs), 2011, 46.

51. Rautava, "Russia's Economic Policy," 45.

52. Rautava, "Russia's Economic Policy," 46.

53. Linda Jakobson, et. al, *China's Energy and Security Relations with Russia: Hopes, Frustrations and Uncertainties* (Stockholm, Sweden: SIPRI), October 2011, accessed October 13, 2013, http://books.sipri.org/files/PP/SIPRIPP29.pdf.

54. Ministry of Defence, Government of India, *Annual Report 2011–2012*, 2012, 195.

55. Ministry of Defence, Government of India, *Annual Report 2011–2012*, 2012, 195.

56. Ministry of Defence, Government of India, *Annual Report 2011–2012*, 2012, 195.

57. Japanese Ministry of Defense, *Defense of Japan 2012*, 2012, 70.

58. Ministry of External Affairs, Government of India, "India–Russia Relations," July 2013, accessed October 13, 2013, http://www.mea.gov.in/Portal/ForeignRelation/India-Russia_Relations.pdf.

59. Ministry of External Affairs, Government of India, "India–Russia Relations," July 2013, accessed October 13, 2013, http://www.mea.gov.in/Portal/ForeignRelation/India-Russia_Relations.pdf.

60. Japanese Ministry of Defense, *Defense of Japan 2012*, 2012, 289.

61. Japanese Ministry of Defense, *Defense of Japan 2012,* 2012, 289.
62. Japanese Ministry of Defense, *Defense of Japan 2012,* 2012, 289.
63. Japanese Ministry of Defense, *Defense of Japan 2012,* 2012, 289.
64. Linda Jakobson, et. al, *China's Energy and Security Relations with Russia.*
65. Linda Jakobson, et. al, *China's Energy and Security Relations with Russia.*
66. Linda Jakobson, et. al, *China's Energy and Security Relations with Russia.*
67. SIPRI Arms Transfers Database, <http://www.sipri.org/databases/armstransfers>.
68. Linda Jakobson, et. al, *China's Energy and Security Relations with Russia.*

8

Western Hemisphere

The Western Hemisphere contains some the largest proven oil reserves in the world but is only a modest oil supplier to China, India, and Japan. In this respect, China, India, and Japan imported respectively 11 percent, 14 percent, and 3 percent in 2012 according to BP. In 2009, they imported respectively 7 percent, 9 percent, and less than 1 percent according to FACTS Global Energy. They intersect at Venezuela, which holds the largest proven oil reserves in the region.

The PRINCE model predicts that China will continue to explore for oil in, and import oil from, the Western Hemisphere based on a probability of 94 percent or greater.[1]

WESTERN HEMISPHERE: POLITICAL IMPLICATIONS FOR INDIA AND JAPAN

China's growing interest in the Western Hemisphere, specifically Latin America and the Caribbean, is a fairly new phenomenon that has developed in recent years.[2] Using soft power and "dollar diplomacy" as its primary diplomatic tools, China's strategic agenda in the region is primarily driven by economic interests such as trade with the strategically economic states of Argentina, Brazil, Chile, and Mexico, although China has also noticeably intensified its diplomatic engagement with the region in recent years because of political competition with Taiwan.[3] In this respect, China desires to politically isolate Taiwan and obtain diplomatic recognition from those countries that currently recognize Taiwan in the Caribbean,

Central America, and South America: El Salvador, Guatemala, Honduras, Nicaragua, Panama, Belize, Dominican Republic, Haiti, St. Kitts and Nevis, St. Lucia, Vincent and the Grenadines, and Paraguay. It should be noted that of the twelve countries in the region that formally recognize Taiwan, only one—Guatemala—sporadically exports oil to China.

China also seeks to strengthen its diplomatic relations with the resource-rich nations of the Western Hemisphere to help feed its resource needs and expanding economy.[4] Countries such as Canada, Colombia, and Venezuela increasingly play important roles to China's energy security as China seeks to find reliable and secure petroleum resources in the region. Mexico could potentially play an important role to China's energy security in the near future if Chinese oil companies are able to acquire licenses to explore for and produce oil. However, Petroleos Mexicanos (the Mexican state-owned oil monopoly) would not be able to increase its crude oil production to meet China's current energy needs.[5]

Cuba will play an important role to China's energy security in the near future because of a recent discovery of 100 million barrels of oil offshore Cuba. In this respect, Cuba's state oil company Cubapetroleo (CUPET) has signed a production sharing agreement with Sinopec after Fidel Castro's announcement of the oil discovery. Such a commercial venture, in conjunction with China's established trade/intelligence relationship with Cuba, will increase China's presence in Cuba.[6]

Politically, India seeks to further relations with the Western Hemisphere for economic reasons. In particular, Latin America has become a breeding ground for trade and investments across the globe owing to several factors like the opening up of its economy, reduction of tariff/non-tariff barriers, modernization of infrastructure, privatization and liberalization of financial markets and the establishment of democracy in most parts of the region.[7] The potential for bilateral trade is huge as the economies of India and Latin America are almost similar.[8] In addition, India and Latin America, also known as the Community of Latin American and Caribbean States (CELAC), share common democratic values and development aspirations; and have common commitment to strengthening South-South cooperation in the interest of developing countries.[9] CELAC gives India a platform to interact directly with the thirty-three sovereign countries in the region.[10] Most of these countries are favorably disposed towards India.[11]

Politically, Japan conducts diplomacy in CELAC to: strengthen economic relations; support the stable development of the region; and advance cooperation in the international arena.[12]

Politically, Japan supports Japanese companies by establishing economic partnership agreements (EPAs), investment treaties, tax treaties, and other legal frameworks, as well as bilateral consulting.[13] Japan actively promotes the Japanese infrastructure development in CELAC as infrastructural demands in CELAC are expected to increase due to its economic growth.[14] Japan is also working to secure stable supplies of natural resources and food through deepening cooperative relations with CELAC that have abundant natural resources and food.[15]

China's hydrocarbon energy trading with the Western Hemisphere is unlikely to result in diplomatic or political competition for oil energy resources between China and India or China and Japan. China, India, and Japan have good diplomatic relations with most if not all of the countries in the region. Further, crude oil exports from the region to China, India, and Japan are modest compared to the Near East. Lastly, most countries in the region are favorably disposed towards India and Japan because they share common democratic values and development aspirations.

China–Venezuela–India–Japan Diplomatic/Political Relations

Indo-Venezuelan diplomatic relations are cordial and friendly.[16] The fiftieth anniversary of the establishment of diplomatic relations was celebrated in 2009.[17] There is mutual goodwill and cooperation between the two countries including multilateral fora.[18] Both countries are members of the Non-Aligned Movement and the G-15. Further, both share similar views on major international, political, and economic issues.[19]

India and Venezuela have also cooperated on a number of important political issues affecting both of them. In 2005, six agreements and memorandums of understanding (MOUs) were signed including the establishment of an Indo-Venezuelan Joint Commission.[20] The Joint Commission discussed cooperation in areas such as energy, infrastructure and railways, low cost housing, trade, science and technology, culture, poverty alleviation, consumer protection and standards, health, agriculture and defense.[21] In August 2012, foreign ministers met to discuss to respond to regional issues and address international challenges such as including UN reforms, the international banking crisis, climate change, and international terrorism.[22] Both countries also discussed and agreed to work more closely to further strengthen and diversify bilateral energy cooperation.[23]

Japanese–Venezuelan diplomatic relations appear to be close since its establishment in August 1938.[24] Bilateral cooperation is the heart of Japanese–Venezuelan diplomatic relations in the areas of economics, commerce, science, technology, environment, and infrastructure.[25] In terms of energy cooperation, Japan and Venezuela signed a memorandum of understanding (MOU) in March 2009.

Sino-Venezuelan diplomatic relations have developed smoothly since 1974, when formal ties were established. Further, in terms of the levels of political dialogue, consultations and agreements, diplomacy has risen to new heights since 1999 due to numerous high-level official visits.[26] Of particular significance were the six visits that Venezuelan President Hugo Chavez made to China before his demise in 2013 and five reciprocal visits senior Chinese officials made to Venezuela in 2001, 2005, 2007, 2008, and 2009.[27]

The results of these high-level visits have been extraordinary as evidenced by an unprecedented increase in bilateral trade and cooperation.[28] In this respect, 460 bilateral agreements have been signed to date, of which 98 percent were finalized during Hugo Chavez's presidency.[29] Included in these agreements are 160 projects in virtually every field of cooperation.[30] These fields include agriculture, energy, planning, science and technology, telecommunications, mining, industries, manufacturing, infrastructure, aerospace, and homebuilding.[31] Given the high level of cooperation between the two countries, their relationship could be characterized as strategic in nature.

Since Venezuelan President Nicolas Maduro has taken office in April 2013, more than twelve new bilateral agreements have been signed in the fields of energy, construction, science, and technology.[32] In terms of significant bilateral agreements negotiated, Sinopec agreed to invest $14 billion in Venezuela's Orinoco Oil Belt to produce oil jointly with Venezuela's state-owned oil company, PDVSA.[33] Further, China Development Bank agreed to provide $5 billion for social reform and $700 million to develop Venezuela's mining sector in return for oil.[34] Lastly, China's EXIM Bank (Import-Export Bank) agreed to finance the development of a port for Pequiven, a petrochemical company of Venezuela for $391 million.[35]

WESTERN HEMISPHERE: ECONOMIC IMPLICATIONS FOR INDIA AND JAPAN

Although China seeks diplomatic recognition as the legitimate representative of China in international venues—especially with those countries in the Western Hemisphere that formally recognize Taiwan and desires to strengthen its political alliances in international and trade forums, China's primary interest in the Western Hemisphere—especially in Latin America—appears to be economically motivated, with China eager for access to such commodities as iron and other ores, soybeans, copper, iron and steel, integrated circuits and other electrical machinery, and oil to meet the demands of its booming economy and access to regional markets for its manufactured goods.[36] The Western Hemisphere—especially Latin America—is rich in energy, agricultural, and mineral resources.

For the Western Hemisphere, especially Latin America, China is an ideal source for infrastructure investment because many Latin American countries suffer from poverty and indebtedness. In this respect, China has invested $87 billion in Latin America since 2005 according to the Inter-American Dialogue.[37] The largest infrastructure recipients have been Venezuela ($28 billion), Argentina ($11.8 billion), the Bahamas ($2.5 billion), and Mexico ($1 billion).[38] Infrastructure funds were used to develop or expand construction, housing, roads and highways, airports, trains, subways, shorelines, and telecommunications.[39] Thus, many Latin American countries are increasingly viewing China not only as a key trading partner, but also as a potential source of much-needed infrastructure investment.[40] Latin America also economically benefits from trade diversification, foreign direct investment, low-cost imports, and growth in sectors that are complementary to China's trade with the region.[41]

India and Japan's interest in the Western Hemisphere is primarily economic. Both seek to strengthen commercial relations because the economic potential is great. In particular, Latin America and the Caribbean is a new breeding ground for Indian and Japanese trade and investment. Further, both countries seek to secure stable supplies of natural resources and food for their populations. Japan is particularly vulnerable because it is an island nation.

China's oil energy trading and investment opportunities in the Western Hemisphere are unlikely to have any economic implications for India and Japan. India and Japan's oil energy trading and investment footprint is small compared to China especially in Latin America and the Caribbean. Further, although India and Japan's overall trade volume is not as large as China's trade volume in the Western Hemisphere, which includes the United States, Canada, Mexico, Chile, and Venezuela; the Andean Community; Central American Common Market; and Mercosur, it is considerable. Lastly, India and Japan share common democratic values with most if not all of the countries in the Western Hemisphere.

China–Venezuela–India–Japan Economic Relations

Indo-Venezuelan economic relations are generally good. Bilateral trade has increased from $532.7 million to $6.95 billion between 2007 and 2012.[42] Venezuela is India's nineteenth-largest trading partner accounting for 1 percent of India's global trade in 2012.[43] Conversely, India is Venezuela's fourth-largest trading partner accounting for approximately 7 percent of Venezuela's global trade in 2012.[44]

In terms of bilateral trade, India exports to Venezuela pharmaceuticals, chemicals, calcined petroleum coke (CPC), textiles and engineering products such as scooters, equipment, and machinery.[45] Venezuela exports to India crude oil, iron pellets, and electrical cables.[46]

The Joint Commission is the main mechanism for bilateral cooperation including energy cooperation. In this respect, a bilateral agreement on cooperation in the hydrocarbons sector was signed in March 2005, along with an MOU offering ONGC Videsh Ltd. (OVL) opportunities for oil and gas exploration in Venezuela, including a heavy oil field in the Orinoco Belt, following which OVL opened its office in Venezuela.[47] Further, in April 2008, an agreement between OVL and CVP (a subsidiary of PDVSA, Venezuela's national oil company) was signed for the creation of a joint venture called "Petrolera Indo Venezolana SA" for production and exploration of oil in the San Cristobal field in which OVL has a 40 percent stake while PDVSA has the remaining 60 percent share.[48] OVL's investment in the San Cristobal Project was $355.7 million.[49] In February 2010, ONGC Videsh Limited (OVL), Indian Oil Corporation (IOC), Oil India Ltd. (OIL), Repsol of Spain and Petronas of Malaysia formed an international corsortium to develop a multi-billion dollar integrated onshore Carabobo oil project in the Orinoco Belt of Venezuela.[50] The three Indian companies together hold 18 percent; Petronas and Repsol each have 11 percent.[51] PDVSA holds the remaining 60 percent of the stake.[52] An international joint venture, PetroCarabobo S.A., was created in May 2010 from the international consortium and the first oil was produced in December 2012.[53] Moreover, India's Reliance Industries Limited (RIL) signed a fifteen-year crude supply contract with PDVSA, Venezuelan′s State Oil Company, under which Venezuela will export to four hundred thousand barrels of oil per day to India.[54] Despite India's burgeoning oil energy stakes, Venezuela is a modest supplier of crude oil. Venezuela meets approximately 5 percent of India's annual oil requirements.

Japanese–Venezuelan economic relations are modest. Bilateral trade was $1.2 billion in 2012 according to the European Union.[55] Venezuela is Japan's forty-eighth-largest trading partner accounting for less than 1 percent of Japan's global trade in 2012.[56] Conversely, Japan is Venezuela's sixteenth-largest trading partner accounting for less than 1 percent of Venezuela's global trade in 2012.[57]

In terms of bilateral trade, Japan exports to Venezuela mainly automobiles, automotive parts, motorcycles, electronic devices, electrodes and brushes, chassis, vehicles parts and accessories, sound reproduction devices, video and image devices.[58] About 57 percent of Japanese exports to Venezuela are gasoline and diesel engines but the automotive sector represents 80 percent of Japanese total exports to Venezuela.[59] Venezuela exports to Japan mainly non-manufactured products: aluminum, bitumen and natural asphalts, iron, albumens, copper scrap and waste, laminated products, laminated non-alloy steel or iron, iron products, and cacao grains.[60] Approximately 63 percent of Venezuela's exports to Japan are aluminum and aluminum products, 15 percent is iron and iron products and 18 percent is asphalt and/or bitumen.[61]

Currently, Japan does not have oil energy stakes in Venezuela. However, Japan and Venezuela signed a memorandum of understanding (MOU) on energy cooperation in April 2009 in which Japan Oil, Gas and Metals National Corporation (JOGMEC), Inpex Corporation, and Mitsubishi Corporation would pursue a joint feasibility study with Venezuela on the Orinoco Oil Belt.[62] The Orinoco Oil Belt is significant because it is estimated to have the world's second-largest crude oil reserves.[63]

Sino-Venezuelan economic relations are very robust. Bilateral trade has increased from $350 million to $23 billion between 2000 and 2012.[64] Venezuela is China's twenty-sixth-largest trading partner accounting for less than 1 percent of China's global trade in 2012.[65] Conversely, China is Venezuela's second-largest trading partner accounting for 15 percent of Venezuela's global trade in 2012.[66]

In terms of bilateral trade, China exports to Venezuela electromechanical equipment, shoes, textiles, and garments while Venezuela exports to China iron ore and finished oil. China's foreign direct investment in Venezuela has been $44.5 billion since 2005.[67]

In terms of energy trading, China imported approximately 6 percent of its oil from Venezuela in 2012 according to FACTS Global Energy. Further, China has burgeoning investments in Venezuela's upstream sector including equity in at least two exploration blocks. Venezuela has also agreed to build an oil refinery in Guangzhou to accommodate Venezuela's sour oil. See appendix B.

China's oil energy stakes in Venezuela is unlikely to result in either Sino-Indian or Sino-Japanese economic competition for energy resources because India and Japan's oil energy stakes are modest compared to China. Further, Venezuela is a modest supplier of crude oil to India and Japan. In addition, Japan's ally, the United States, exercises considerable political and economic influence given that it is the largest importer and refiner of Venezuela's heavy crude oil. Further, China is not technologically equipped to refine Venezuela's heavy crude oil. Special refineries are required to refine Venezuela's heavy, sulfur-rich crude petroleum which China does not have.[68] Thus, most of the crude petroleum that China currently buys from Venezuela gets traded in Singapore to third markets.[69]

WESTERN HEMISPHERE: MILITARY IMPLICATIONS FOR INDIA AND JAPAN

China is rapidly emerging as a military player in the region.[70] According to CSIS, "China is using military diplomacy, educational exchanges and medical and technical assistance in a measured way to support its broader political and economic interests in Latin America."[71] Obviously, greater access to natural resources such as oil fits into the broader political and economic scheme of China's foreign relations in the Western Hemisphere and will keep the Chinese military engaged in the region as part of a comprehensive effort to build mutually beneficial economic development, trade, military links, and political relations rooted in principles of "South-South Cooperation" similar to China's efforts in Africa. Thus, the Chinese People's Liberation Army (PLA) in reality operates increasingly as an instrument of diplomatic statecraft of the People's Republic of China.[72] The PLA's role "constitutes just one part of an expanding presence that China manifests as a 'major power' on the global stage."[73] Further, "the PLA's involvement in Latin America illustrates China's pursuit of a multi-faceted strategy to expand its global presence."[74]

Chinese military engagement within the region is best understood in terms of five interrelated types of activities: meetings between senior military officials; lower-level military-to-military interactions; military sales; military-relevant commercial interactions; and Chinese physical presence within Latin America with military-strategic implications.[75]

In terms of military presence in the Western Hemisphere, China has nine defense attaché offices in host countries which have proven oil reserves: Argentina, Belize, Brazil, Chile, Colombia, Cuba, Ecuador, Mexico, Peru, and Uruguay. Of these, only four do not produce oil for China: Belize, Chile, Mexico, and Uruguay. In addition to these official links, the Chinese military has been engaged in Bolivia, Ecuador, Peru, and Venezuela through high-level military visits and arms sales.

India's military engagement of the Western Hemisphere has been limited to Brazil, Ecuador, and the United States. India and Brazil conducted a high-level joint defense meeting in October 2011.[76] Moreover, India and Ecuador signed a memorandum of understanding (MOU) on defense cooperation in March 2011.[77] Lastly, India and the United States increased bilateral defense cooperation through military cooperation activities, expert exchanges, high-level visits, defense research, and joint exercises.[78] Japan's military engagement of the Western Hemisphere has been limited but strategic. In this respect, Japan and the United States have a formal military alliance, bilateral joint training, defense exchanges, and U.S. military forces stationed in and around Japan. Japan has also conducted high-level military and defense ministerial exchanges with Canada in 2010 and 2011.[79]

China–Venezuela–India–Japan Military Relations

Indo-Venezuelan military relations are unknown. According to the Indian Ministry of Defence, India and Venezuela do not have any bilateral military agreements, defense cooperation, or military exchanges.

Japanese–Venezuelan military relations are also unknown. According to the Japanese Ministry of Defense, Japan and Venezuela do not have any bilateral military agreements, defense cooperation, or military exchanges.

Sino-Venezuelan military relations are very good. Bilateral military ties include military and defense-related educational exchanges, frequent official visits by senior defense leaders, cultural events such as visits to Venezuela by PLA band and acrobatic troupes, and intelligence cooperation. In terms of arms sales, China has supplied Venezuela: ten JY-11 air search radars, three JYL-1 air search radars, eighteen K-8 basic trainers and ground attack aircraft, eight Y-8 transport aircraft, and one hundred PL-5E short-range air-to-air missiles according to SIPRI.[80]

China's energy stakes in Venezuela is unlikely to result in either Chinese and Indian or Chinese and Japanese military competition for oil energy resources because the United States is Venezuela's largest petroleum customer in terms of physical volume and sales. Further, only the United States can refine Venezuela's heavy crude oil. The fact that the United States is Venezuela's largest oil customer and the only country that can process Venezuela's heavy crude oil makes Venezuela economically dependent on the United States. In addition, China does not wish to damage its overall relations with the United States by appearing to interfere in America's traditional sphere of political influence—the Western Hemisphere. Also, Venezuela is a modest supplier of crude petroleum to China, India, and Japan. Lastly, most of China's oil purchases from Venezuela are traded in Singapore to third markets.[81]

NOTES

1. Prince and Panda website, "PRINCE_SCORES.pdf," accessed October 17, 2013, http://www.princeandpanda.com.
2. U.S. Library of Congress, Congressional Research Service, *China's Foreign Policy and "Soft Power" in South America, Asia, and Africa: A Study Prepared for the Committee on Foreign Relations United States Senate* (Washington, DC: U.S. Government Printing Office, April 2008), 16.
3. Riordan Roett and Guadalupe Paz, "Introduction: Assessing the Implications of China's Growing Presence in the Western Hemisphere," in Riordan Roett and Guadalupe Paz, eds., *China's Expansion Into The Western Hemisphere: Implications for Latin America and the United States.* (Washington, DC: Brookings Institution Press, 2008), 16.
4. U.S. Library of Congress, Congressional Research Service, *Comparing Global Influence: China's and U.S. Diplomacy, Foreign Aid, Trade, and Investment in the Developing World* (Washington, DC: U.S. Government Printing Office), RL 34620, August 15, 2008, 158.

5. Chung-Chian Teng, "Hegemony or Partnership: China's Strategy and Diplomacy Toward Latin America," in Joshua Eisenman, Eric Heginbotham, and Derek Mitchell, eds., *China and the Developing World: Beijing's Strategy for the Twenty-First Century*. (Armonk, NY: M.E. Sharpe, 2007), 97.

6. U.S. House, Subcommittee on the Western Hemisphere, *China's Influence on the Western Hemisphere*, Hearing on April 6, 2005, 109th Congress, Session 1, 2005, 55.

7. Rupa Chanda, Debasmita Panja, and Swapneela Biswas, "India–Latin America Trade Relationship," *Tejas*, July 2012, accessed October 13, 2013, http://tejas.iimb.ac.in/ articles/102.php.

8. Chanda, Panja, and Biswas, "India–Latin America Trade Relationship."

9. Ministry of External Affairs, Government of India, "Community of Latin American and Caribbean States (CELAC)," August 2012, accessed October 13, 2013, http://www.mea.gov.in/Portal/ForeignRelation/celac-august-2012.pdf.

10. Ministry of External Affairs, Government of India, "Community of Latin American and Caribbean States (CELAC)," August 2012, accessed October 13, 2013, http://www.mea.gov.in/Portal/ForeignRelation/celac-august-2012.pdf.

11. Ministry of External Affairs, Government of India, "Community of Latin American and Caribbean States (CELAC)," August 2012, accessed October 13, 2013, http://www.mea.gov.in/Portal/ForeignRelation/celac-august-2012.pdf.

12. Ministry of Foreign Affairs, Japan, "Japan's Foreign Policy by Region," in *Diplomatic Bluebook 2012* (Tokyo, Japan), April 2012, 24.

13. Ministry of Foreign Affairs, Japan, "Japan's Foreign Policy by Region," in *Diplomatic Bluebook 2012* (Tokyo, Japan), April 2012, 24.

14. Ministry of Foreign Affairs, Japan, "Japan's Foreign Policy by Region," in *Diplomatic Bluebook 2012* (Tokyo, Japan), April 2012, 24.

15. Ministry of Foreign Affairs, Japan, "Japan's Foreign Policy by Region," in *Diplomatic Bluebook 2012* (Tokyo, Japan), April 2012, 24.

16. Embassy of India to Venezuela website, "India–Venezuela Relations," March 2012, accessed October 13, 2013, http://www.embindia.org/India-Venezuela_Rel.php.

17. Ministry of External Affairs, Government of India, "India–Venezuela Relations," June 2013, accessed October 13, 2013, http://www.mea.gov.in/Portal/ForeignRelation/ Venezuela_June__2013.pdf.

18. Ministry of External Affairs, Government of India, "India–Venezuela Relations," June 2013, accessed October 13, 2013, http://www.mea.gov.in/Portal/ForeignRelation/ Venezuela_June__2013.pdf.

19. Ministry of External Affairs, Government of India, "India–Venezuela Relations," June 2013, accessed October 13, 2013, http://www.mea.gov.in/Portal/ForeignRelation/ Venezuela_June__2013.pdf.

20. Ministry of External Affairs, Government of India, "India–Venezuela Relations," June 2013, accessed October 13, 2013, http://www.mea.gov.in/Portal/ForeignRelation/ Venezuela_June__2013.pdf.

21. Embassy of India to Venezuela website, "India–Venezuela Relations," March 2012, accessed October 13, 2013, http://www.embindia.org/India-Venezuela_Rel.php.

22. Ministry of External Affairs, Government of India, "India–Venezuela Relations," June 2013, accessed October 13, 2013, http://www.mea.gov.in/Portal/ForeignRelation/ Venezuela_June__2013.pdf.
23. Ministry of External Affairs, Government of India, "India–Venezuela Relations," June 2013, accessed October 13, 2013, http://www.mea.gov.in/Portal/ForeignRelation/ Venezuela_June__2013.pdf.
24. Embassy of Venezuela to Japan website, "Bilateral Relations," accessed October 13, 2013, http://www.venezuela.or.jp/content/category/11/37/95/.
25. Embassy of Venezuela to Japan website, "Bilateral Relations," accessed October 13, 2013, http://www.venezuela.or.jp/content/category/11/37/95/.
26. "Venezuela, China: 38 years of diplomatic relations," *ChinaDaily.com.cn*, June 28, 2012, accessed October 13, 2013, http://www.chinadaily.com.cn/cndy/2012-06/28/content_15528289.htm.
27. "Venezuela, China: 38 years of diplomatic relations," *ChinaDaily.com.cn*, June 28, 2012, accessed October 13, 2013, http://www.chinadaily.com.cn/cndy/2012-06/28/content_15528289.htm.
28. "Venezuela, China: 38 years of diplomatic relations," *ChinaDaily.com.cn*, June 28, 2012, accessed October 13, 2013, http://www.chinadaily.com.cn/cndy/2012-06/28/content_15528289.htm.
29. "Venezuela, China: 38 years of diplomatic relations," *ChinaDaily.com.cn*, June 28, 2012, accessed October 13, 2013, http://www.chinadaily.com.cn/cndy/2012-06/28/content_15528289.htm.
30. "Venezuela, China: 38 years of diplomatic relations," *ChinaDaily.com.cn*, June 28, 2012, accessed October 13, 2013, http://www.chinadaily.com.cn/cndy/2012-06/28/content_15528289.htm.
31. "Venezuela, China: 38 years of diplomatic relations," *ChinaDaily.com.cn*, June 28, 2012, accessed October 13, 2013, http://www.chinadaily.com.cn/cndy/2012-06/28/content_15528289.htm.
32. Yagmur Ersan, "China and Venezuela strengthen bilateral relations," *Turkish Weekly*, October 2, 2013, accessed October 13, 2013, http://www.turkishweekly.net/ news/157211/china-and-venezuela-strengthen-bilateral-relations.html.
33. Ersan, "China and Venezuela strengthen bilateral relations."
34. Ersan, "China and Venezuela strengthen bilateral relations."
35. Ersan, "China and Venezuela strengthen bilateral relations."
36. Catherine E. Bliss, "China's Projection of Soft Power in the Americas", in Carola McGiffert, ed., *Chinese Soft Power and Its Implications for the United States* (Washington, DC: CSIS Press), March 2009, 45 and the U.S. Library of Congress, *China's Foreign Policy and "Soft Power" in South America, Asia, and Africa*, 16.
37. Inter-American Dialogue, "China–Latin America Finance Database," accessed October 13, 2013, http://thedialogue.org/map_list.
38. Inter-American Dialogue, "China–Latin America Finance Database," accessed October 13, 2013, http://thedialogue.org/map_list.
39. Inter-American Dialogue, "China–Latin America Finance Database," accessed October 13, 2013, http://thedialogue.org/map_list.
40. Xiang, "An Alternative Chinese View," 55.

41. Roett and Paz, "Introduction," 16.

42. Ministry of External Affairs, Government of India, "India–Venezuela Relations," June 2013, accessed October 13, 2013, http://www.mea.gov.in/Portal/ForeignRelation/ Venezuela_June__2013.pdf.

43. European Commission Directorate-General for Trade, "India: EU Bilateral Trade and Trade with the World," July 5, 2013, accessed October 13, 2013, http://trade.ec.europa.eu/doclib/docs/2006/ september/tradoc_113390.pdf.

44. European Commission Directorate-General for Trade, "Venezuela: EU Bilateral Trade and Trade with the World," July 5, 2013, accessed October 13, 2013, http://trade.ec.europa.eu/doclib/docs/2006/ september/tradoc_113462.pdf.

45. Ministry of External Affairs, Government of India, "India–Venezuela Relations," June 2013, accessed October 13, 2013, http://www.mea.gov.in/Portal/ForeignRelation/ Venezuela_June__2013.pdf.

46. Ministry of External Affairs, Government of India, "India–Venezuela Relations," June 2013, accessed October 13, 2013, http://www.mea.gov.in/Portal/ForeignRelation/ Venezuela_June__2013.pdf.

47. Embassy of India to Venezuela website, "India–Venezuela Relations," March 2012, accessed October 13, 2013, http://www.embindia.org/India-Venezuela_Rel.php.

48. Embassy of India to Venezuela website, "India–Venezuela Relations," March 2012, accessed October 13, 2013, http://www.embindia.org/India-Venezuela_Rel.php.

49. Embassy of India to Venezuela website, "India–Venezuela Relations," March 2012, accessed October 13, 2013, http://www.embindia.org/India-Venezuela_Rel.php.

50. Embassy of India to Venezuela website, "India–Venezuela Relations," March 2012, accessed October 13, 2013, http://www.embindia.org/India-Venezuela_Rel.php.

51. Embassy of India to Venezuela website, "India–Venezuela Relations," March 2012, accessed October 13, 2013, http://www.embindia.org/India-Venezuela_Rel.php.

52. Embassy of India to Venezuela website, "India–Venezuela Relations," March 2012, accessed October 13, 2013, http://www.embindia.org/India-Venezuela_Rel.php.

53. Ministry of External Affairs, Government of India, "India–Venezuela Relations," June 2013, accessed October 13, 2013, http://www.mea.gov.in/Portal/ForeignRelation/ Venezuela_June__2013.pdf.

54. Ministry of External Affairs, Government of India, "India–Venezuela Relations," June 2013, accessed October 13, 2013, http://www.mea.gov.in/Portal/ForeignRelation/ Venezuela_June__2013.pdf.

55. European Commission Directorate-General for Trade, "Venezuela: EU Bilateral Trade and Trade with the World," July 5, 2013, accessed October 13, 2013, http://trade.ec.europa.eu/doclib/docs/2006/ september/tradoc_113462.pdf.

56. European Commission Directorate-General for Trade, "Japan: EU Bilateral Trade and Trade with the World," July 5, 2013, accessed October 13, 2013, http://trade.ec.europa.eu/doclib/docs/2006/ september/tradoc_113403.pdf.

57. European Commission Directorate-General for Trade, "Venezuela: EU Bilateral Trade and Trade with the World," July 5, 2013, accessed October 13, 2013, http://trade.ec.europa.eu/doclib/docs/2006/ september/tradoc_113462.pdf.

58. Embassy of Venezuela to Japan website, "Commercial Relations," accessed October 13, 2013, http://www.venezuela.or.jp/content/view/83/95/.

59. Embassy of Venezuela to Japan website, "Commercial Relations," accessed October 13, 2013, http://www.venezuela.or.jp/content/view/83/95/.

60. Embassy of Venezuela to Japan website, "Commercial Relations," accessed October 13, 2013, http://www.venezuela.or.jp/content/view/83/95/.

61. Embassy of Venezuela to Japan website, "Commercial Relations," accessed October 13, 2013, http://www.venezuela.or.jp/content/view/83/95/.

62. JOGMEC, "JOGMEC participates in an extra heavy oil development project of the Orinoco Oil Belt in the Bolivarian Republic of Venezuela," news release, April 12, 2010, accessed October 13, 2013, http://www.jogmec.go.jp/english/news/release/release0048. html.

63. James Suggett, "Japan and Venezuela Initiate Joint Orinoco Oil Projects and Expand Economic Impacts," *Venezuelaanalysis.com,* April 6, 2009, accessed October 13, 2013, http://venezuelanalysis.com/news/4352.

64. Jiao Wu, "China signs 12 deals with Venezuela," *ChinaDaily.com.cn,* September 23, 2013, accessed October 13, 2013, http://www.chinadaily.com.cn/world/2013-09/23/content_16985752.htm.

65. European Commission Directorate-General for Trade, "China: EU Bilateral Trade and Trade with the World," July 5, 2013, accessed October 13, 2013, http://trade.ec.europa.eu/doclib/docs/2006/ september/tradoc_113366.pdf.

66. European Commission Directorate-General for Trade, "Venezuela: EU Bilateral Trade and Trade with the World," July 5, 2013, accessed October 13, 2013, http://trade.ec.europa.eu/doclib/docs/2006/ september/tradoc_113462.pdf.

67. Inter-American Dialogue, "China–Latin America Finance Database," accessed October 13, 2013, http://thedialogue.org/map_list.

68. Javier Corrales, "China and Venezuela's Search for Oil Markets," in Alex E. Fernandez Jilberto and Barbara Hogenboom, eds., *Latin America Facing China: South-South Relations beyond the Washington Consensus.* (New York, NY: Berghahn Books, 2012), 124.

69. Corrales, "China and Venezuela's Search for Oil Markets," 124.

70. Loro Horta, "In Uncle Sam's Backyard: China's Military Influence in Latin America," *Military Review,* September–October 2008, 51.

71. Stephen J. Flanagan, "China's Military Diplomacy and Engagement in Latin America," in Carola McGiffert, ed., *Chinese Soft Power and Its Implications for the United States* (Washington, DC: CSIS Press, March 2009), 56.

72. Cynthia Watson, "The PLA in Latin America," *China Brief,* (Jamestown Foundation), Volume VII, Issue 20, October 31, 2007, 1.

73. Watson, "The PLA in Latin America," 1.

74. Watson, "The PLA in Latin America," 1.

75. R. Evan Ellis, *China–Latin America Military Engagement: Good Will, Good Business and Strategic Position* (Carlisle, PA: Strategic Studies Institute), August 2011, 10.

76. Ministry of Defence, Government of India, *Annual Report 2011–2012*, 2012, 198.

77. Ministry of Defence, Government of India, *Annual Report 2011–2012*, 2012, 198.

78. Ministry of Defence, Government of India, *Annual Report 2011–2012*, 2012, 196–198.

79. Japanese Ministry of Defense, *Defense of Japan 2012*, 2012, 279.

80. SIPRI Arms Transfers Database, <http://www.sipri.org/databases/armstransfers>.

81. Corrales, "China and Venezuela's Search for Oil Markets," 124.

9

+

Summary

Comprehensive regional analysis indicates that Sub-Saharan Africa and the Near East are the two most unstable regions of the world. This presents a vexing problem for China because 77 percent of its oil is imported from these regions combined, of which 100 percent is transported by sea along the vital sea lines of communication (SLOCs), the commercial shipping routes that the United States Navy and allied navies patrol constantly. Even though there are few conditions that are likely to bring a military confrontation between China, India, and Japan over oil energy resources, it is not inconceivable that the United States may need to come to their assistance in the near future to keep open some of the sea lines of communication such as the Bab-el-Mandeb on the Red Sea or the Strait of Hormuz in the Persian Gulf in light of the fact that the bulk of their imported oil passes through these vital chokepoints. What must also be taken in consideration is that Japan, America's foremost ally in Asia, imports a considerable amount of its oil from the Near East—approximately 78 percent—according to the U.S. Energy Information Administration. That fact may generate additional pressure on the United States to intervene in any conflict affecting the security of the chokepoints.

Through the adroit use of soft power, China's strategic engagement in Sub-Saharan Africa for the pursuit of scarce energy resources potentially undermines India and Japan's economic efforts to obtain reliable crude oil and petroleum sources. Cases in point are Angola, Nigeria, and Sudan. Further, India and Japan will likely face economic competition in the Near East as energy resources become scarce, in the absence of fuel-saving technology alternatives, and strategic partnerships with Iran, Saudi

Arabia, and the UAE. It is unlikely India and Japan will face economic competition from China in the near future. However, if the United States becomes less strategically engaged in the Near East, China will likely fill the strategic void and economic competition for energy resources will result. Much will depend on whether the United States stays engaged.

China's hard power and geopolitical influence is growing steadily in tandem with its pursuit and development of scarce energy resources. Chinese warships are currently patrolling the Bab-el-Mandeb in support of international antipiracy operations. China is also Iran's largest conventional weapons supplier and plays a major role in Iran's military modernization and nuclear development program. China has quietly engaged in a strategy of military and economic cooperation with Pakistan, Myanmar, Bangladesh, Sri Lanka, and the Maldives in the Indian Ocean region in order to build and/or have access to ports and airbases to support not only a future tanker fleet but also a blue-water navy capable of projecting power not only in the Indian Ocean region but in the Persian Gulf. Beyond the Indian Ocean region China has been offered the use of a former U.S. airbase in Ecuador.

Overall, China's strategic engagement and rising geopolitical influence in tandem with its adroit pursuit and development of scarce energy resources causes it to be a strategic energy resource competitor economically to India and Japan in those countries where they obtain their petroleum products. China will alter the status quo within the next fifty to one hundred years and surpass the United States as a global power unless it can find a way to reduce its energy dependence on hydrocarbon-based fuels. China has already surpassed India and Japan in terms of crude oil trading and petroleum energy consumption.

II

THE SURPLUS SCENARIOS

10

Surplus Scenarios

BACKGROUND

After decades of importing oil from Central Asia, the Near East, Sub-Saharan Africa, Western Europe, and the Western Hemisphere, China has employed motor vehicle fuel-saving technologies to sustain economic growth and development. These modern technologies consist of hybrid-electric vehicles (HEVs), compressed natural gas (CNG) vehicles and electric vehicles for public and private transportation. In this scenario the major oil consumers (the European Union, India, Japan, and the United States) adopt public and private transportation fuel-saving technologies and are importing considerably less oil in an effort to reduce their dependency on imported oil. It would not be realistic to expect China to adopt transportation fuel-saving technologies if other large oil consumers did not.

As a result of implementing fuel-saving technologies in motor vehicles for public and private transportation, China has cut its annual oil consumption by more than 50 percent. Consequently, global oil prices have dropped, oil consumption has been reduced and the life of global oil supplies has been extended.

The following two scenarios analyze energy futures in which China relies heavily on fuel-saving policies and technologies (hybrid, electric, compressed natural gas) to decrease consumption and increase production. In the competitive surplus scenario, competition (unilateral exploration) exists in South China Sea for oil energy resources while in the cooperative surplus scenario, cooperation (joint exploration) exists in the South China Sea for oil energy resources.

COMPETITIVE SURPLUS SCENARIO

Although China uses considerably less hydrocarbon-based fuels, China still explores for oil both onshore and offshore including the East China Sea and South China Sea. China's unilateral oil exploration and production operations (upstream activities) in the East China Sea and South China Sea has caused considerable consternation with the governments of Brunei, Japan, Malaysia, the Philippines, Taiwan, and Vietnam because of disputed sovereignty claims.

COOPERATIVE SURPLUS SCENARIO

This scenario is different from the competitive scenario in that China jointly explores for oil in the South China Sea and the Gulf of Mexico. Although China uses considerably less hydrocarbon-based fuels, it still explores for oil both onshore and offshore mainland China including the South China Sea. China's efforts to conduct joint oil exploration and production operations (upstream activities) in the South China Sea have improved bilateral relations somewhat with Brunei, Malaysia, the Philippines, Taiwan, and Vietnam.

China has also pursued joint oil exploration and production operations in the Gulf of Mexico with independent oil companies such as ExxonMobil and ChevronTexaco.

PROBABILITY OF OCCURRENCE

As in the competitive dependency scenarios analyzed in chapters 2–8, in the surplus scenarios, a PRINCE score of greater than 60 percent indicates that future events/issues will occur or will likely continue if already in progress. A PRINCE score of 40 to 60 percent means that it is uncertain as to whether the events or issues will occur. Moreover, if any of the aforementioned issues or events has a PRINCE score of less than 40 percent, then they will not occur. See chapter 1 for a description of the PRINCE method. The overall probability of occurrence in the competitive surplus scenario is 77 percent, and the overall probability of occurrence for the cooperative surplus scenario is 54 percent, so the competitive surplus scenario is probable depending on the exact sequence of events and decisions taken while the cooperative surplus scenario is uncertain.[1]

PRINCE scores were calibrated to improve forecasting accuracy by first averaging the PRINCE scores of China's fuel saving technology operations as one group and the PRINCE scores for China's domestic crude oil and petroleum operations as one group. These PRINCE scores were then averaged with the PRINCE scores for each regional oil prospecting operation in the South China Sea (Spratly Islands and the Paracel Islands) and international oil prospecting operation (either the East China Sea or Gulf of Mexico) to get the overall probability of occurrence for each scenario.

MACRO POLITICAL AND ECONOMIC STRATEGIC IMPLICATIONS FOR INDIA AND JAPAN

China's use of alternative fuel technology (hybrid electric vehicles, electric cars, and compressed natural gas cars in addition to gas-powered vehicles) in lieu of fossil fuel technology for public and private transportation will undoubtedly provide for better diplomatic relations with India and Japan with respect to global energy security and cooperation because China is no longer relying on primarily fossil fuels to sustain national economic growth and development. Essentially, China is not tapping into the global oil market. India and Japan will not be energy insecure with respect to China because it faces far less competition from China in acquiring global oil energy resources. But because India and Japan adopted fuel-saving technologies, like China, it also no longer depends on foreign oil.

China's policy of using alternative fuel technologies will not reduce, change, or affect any global political issues between China, India, and Japan that are unrelated to oil energy matters or energy security such as territorial sovereignty issues. China will continue to build and improve its diplomatic relationships with the developing and developed world using its non-ideological, pragmatic approach to maintain its domestic social stability; to ensure the political longevity of the Chinese Communist Party; to sustain its domestic economic growth and development; to increase its foreign trade; and to improve its global political position.

Macro Political and Economic Implications: Competitive Surplus Scenario

China's unilateral exploration, discovery, extraction, and refinement of crude oil in Japan's Exclusive Economic Zone (EEZ), in the Paracel Islands, and in the Spratly Islands may make the East China Sea and South China Sea politically unstable and disrupt the flow of international shipping traffic (including the transit of oil tankers) in the region. This is due in part to greater Chinese military presence in the East China Sea and South China Sea to protect its claimed oil interests in Paracel Islands and in the Spratly Islands. Second, there is an increased probability of armed conflict between China, Japan, Brunei, Malaysia, the Philippines, Taiwan, and Vietnam and possibly the United States over conflicting sovereignty claims in Japan's Exclusive Economic Zone, and in the Paracel and Spratly Islands. This would most likely lead to a severe economic downturn, increasing the probability of political instability and the specter of war in the East China Sea and the South China Sea, resulting from the disruption of crude oil and petroleum exports to Japan and the United States. Increasing brinksmanship between the United States and China may further exacerbate regional tensions.

Macro Political, Economic, and Military Implications: Cooperative Surplus Scenario

Unlike the competitive surplus scenario, China's cooperative efforts to conduct joint exploration, discovery, extraction, and refinement of crude oil in the Paracel Islands and in the Spratly Islands may make South China Sea more politically stable and may improve the flow of international shipping traffic (including the transit of oil tankers) in the region in the absence of sovereignty disputes and a kinetic military exchange. However, its joint operations with independent oil companies in the Gulf of Mexico adjacent to Cuba will not make for greater bilateral diplomatic relations between China and India or China and Japan because oil operations are inconsequential in this region.

China's operations in the South China Sea and East China Sea may make the region more economically stable if joint operations are successful and provide for better trade relations between China and ASEAN. China's cooperative efforts with independent oil companies in the Gulf of Mexico adjacent to Cuba will not necessarily make for greater bilateral economic and trade relations between China and India or China or Japan.

MACRO-ECONOMIC IMPLICATIONS

China's use of alternative fuel technology (hybrid-electric vehicles, electric cars, and compressed natural gas cars in addition to gas-powered vehicles) in lieu of fossil fuel technology for public and private transportation will undoubtedly provide for better economic and trade relations with India and Japan, especially in motor vehicle sales, because of sharing and adoption of motor vehicle fuel-saving technologies between China, India, and Japan. The development and effective use of fuel-saving technologies in China will lower global oil prices, reduce oil consumption, and extend the life of available oil reserves. Yet the success of fuel-saving technologies in China will largely depend on China's actual GDP growth rate and whether the international automotive manufacturing industry can capture China's domestic car sales market. Table 10.1 provides an estimate of GDP growth versus the number of motor vehicles in China in the near future, and table 10.2 provides fuel economy comparisons for popular vehicles in 2013. Consumers could cut up to one-third of their annual fuel cost by driving a fuel-saving technology vehicle.

Table 10.1 National Vehicle Fleet Projections for Three GDP Growth Rates, China

GDP Growth	Year	Cars (millions of vehicles)	Motor Vehicles (millions of vehicles)
10 percent	2015	24.4	68.4
10 percent	2020	43.1	110.2
8 percent	2015	18.0	52.9
8 percent	2020	28.5	77.8
6 percent	2015	13.2	40.7
6 percent	2020	18.7	54.5

Source: Calculated by G. Ingram based on data in China State Statistical Bureau (2002).

Table 10.2 Fuel Economy Comparisons for Popular Vehicles in 2013 Based on 15,000 Miles of Annual Traveling and a Fuel Cost of $3.33 per Gallon

Make (2013)	Model	MPG-City	MPG-Highway	Annual Fuel Cost
Civic	Honda	22	31	$2,000
Civic-Hybrid	Honda	44	44	$1,150
Civic-CNG	Honda	27	38	$1,000
Accord	Honda	24	34	$1,800
Camry	Toyota	25	35	$1,800
Camry-Hybrid	Toyota	43	39	$1,200
Prius	Toyota	51	48	$1,000

Source: *Fuel Economy Guide 2013*, www.fueleconomy.gov.

China will not be deterred from either supporting pariah states like Burma, Cuba, Iran, and Sudan or prevent her from complying with international security and stability protocols on human rights, arms control (including weapons of mass destruction technology transfers). Beijing will continue to build and improve its economic relationships (like its diplomatic relationships) with the developing and developed world using a non-ideological, pragmatic approach to maintain its domestic and international objectives.

MACRO MILITARY IMPLICATIONS

China's use of alternative fuel technology will not necessarily make for better military relations with India or Japan in spite of potentially lower global oil prices, greater diplomatic opportunities, and trade possibilities in commercial vehicle alternative fuel technology between China, Japan, and the United States because of the lack of transparency regarding China's military capabilities and modernization; and China's continued deployment of military personnel around the world in support of its "go-out strategy." India and Japan will not necessarily be more energy secure because of China's military secrecy, modernization of its military forces and growing military relations with the developing and developed world.

REGIONAL ANALYSES

In the surplus scenarios, an analysis of China's oil energy activities and its implications for India and Japan are made for East Asia and Southeast Asia. Further, the Gulf of Mexico is individually addressed in the cooperative surplus scenario. Realistically, the impact of China's oil energy activities on India and Japan will vary (positively, negatively, or none) based on the extent of overall Chinese, Indian, and Japanese political, economic/trade, and military relations, aid and influence, and amount of oil energy resources available in the region. (See appendix A for available oil resources). Available hydrocarbon resources, proven crude oil reserves as of October 5, 2013 are shown below (see table 10.3) according to U.S. Energy Information Administration.

Although the Near East and the Western Hemisphere hold the majority of proven oil reserves and are clearly the most important regions of the world in terms of energy security, these regions are not vital to China's energy security because China uses fuel-saving technologies to sustain economic growth and development.

Table 10.3 Proven Oil Reserves

Region	Billion Barrels	% of World Total
Russian Federation	80	5
Asia-Pacific	47.2	3
Europe and Eurasia	50.9	3
Near East (Middle East & North Africa)	867	54
Africa (Sub-Saharan Africa)	62.5	4
Western Hemisphere (North & South America)	509.2	31
Total	1,616.8	100

Source: U.S. Energy Information Administration, 2013.

EAST ASIA IN THE COMPETITIVE AND COOPERATIVE SURPLUS FUTURES

The PRINCE model predicts that China will continue to prospect for oil in Japan's Exclusive Economic Zone by 71 percent. China will continue to prospect for oil both onshore in the Tarim Basin by 91 percent, in Bohai Bay by 96 percent and coastally by 98 percent. China will also develop and use fuel-saving technology vehicles by 97 percent.[2]

East Asia: Political, Economic, and Military Implications for India and Japan

China's oil energy activities in East Asia are unlikely to shift the balance of political power from either India or Japan to China because of the strong diplomatic relationships that exist between India, Japan, and East Asia nor does oil trading shape or influence political relationships in East Asia.

The efficient development and effective use of fuel-saving technologies in China's motor vehicles will improve Sino-Indian and Sino-Japanese energy security measures because China will no longer rely solely on fossil fuel energy resources to meet their domestic oil energy needs. In this sense, there would be no need for China to import as well as explore for and produce foreign oil (apart from the East China Sea and the South China Sea) because Beijing will have implemented public policies to transform both the public and private transportation industry using fuel-saving technologies. Although China will not have to compete with India and Japan for global oil energy resources, bilateral relations outside of energy security matters may not improve over current conditions. Contentious foreign policy issues like territorial sovereignty issues will remain.

China's hydrocarbon energy activities in East Asia are unlikely to shift the balance of economic power from either India or Japan to China because of the strong economic relationships that exist (aside from oil energy trading) between them. In this respect, oil trading does not play a pivotal role in shaping or influencing economic relations between them. International energy trade in East Asia enhances regional stability as well as political and military relations, although bilateral trade relations in other commodities remain unaffected.

Domestically, the adoption and use of fuel-saving technologies will result in greater fuel economy within China as well as a reduction in annual fuel cost for individual consumers. Table 10.2 provides a comparison of popular vehicle makes and models for the year 2013 to show how overall fuel economy will improve and how annual fuel costs (per unit) will be reduced when using fuel-saving technologies in motor vehicles.

China's hydrocarbon energy activities in East Asia are unlikely to shift the balance of *military power* from either India or Japan to China or lead to a military competition for energy resources because of the close military relationship that exists between the United States, India, and Japan. In this respect, hydrocarbon energy trading does not play a role in shaping or influencing military relations between them, and may not improve bilateral military relations in other areas. China's military modernization, its development of a blue-water navy, and increased global military-to-military ties will occur independently of the development and use of fuel-saving technologies in the public and private transportation sectors.

The Political, Economic, and Military Implications of China's Domestic Hydrocarbon Operations

Domestic hydrocarbon operations (exploration, discovery, extraction and refinement) in territorial China such as the Tarim Basin in Xinjiang, Bohai Bay, and coastal China are unlikely to have any adverse diplomatic, economic, or military consequences for India and Japan. These operations are sovereign in nature and do not involve India, Japan, or any other country except in a capacity where technologically advanced countries like Japan and the United States may be able to provide public and private technical expertise to assist China in producing its natural hydrocarbon energy resources from underground and below the ocean floor. Should Beijing seek greater cooperation from Japan or the United

States all will benefit on many levels, such as increasing their strategic petroleum reserves (SPR) as well as lowering global petroleum prices. Therefore, it behooves China, Japan, and the United States to engage each other and work constructively.

SOUTHEAST ASIA IN A COMPETITIVE SURPLUS FUTURE

The PRINCE model predicts that China will continue to unilaterally prospect for oil in the Paracel Islands by 67 percent. China will continue to unilaterally prospect for oil in the Spratly Islands by 58 percent.[3]

The Political, Economic, and Military Implications of South China Sea Sovereignty Dispute

There is a 67 percent chance that China will continue to extract oil from the Paracel Islands and a 58 percent chance that it will continue operations in the Spratly Islands.[4] Disputed sovereignty claims in the South China Sea over the Paracel Islands (between China, Taiwan, and Vietnam) and Spratly Islands (between China, Malaysia, the Philippines, Taiwan, and Vietnam) could draw the United States inexorably into an accidental war with China in the absence of preventive diplomatic measures. War would not only disrupt trade and commerce throughout Southeast Asia but the Asian Pacific as well, which could have first and second order effects on world trade. China's economy could potentially be undermined leading to major social unrest, and resulting in the overthrow of the Chinese Communist Party.

SOUTHEAST ASIA IN A COOPERATIVE SURPLUS FUTURE

The PRINCE model predicts that China will not jointly prospect for oil in the Paracel Islands with Taiwan and Vietnam. The model also predicts that China will not jointly prospect for oil in the Spratly Islands with Brunei, Malaysia, the Philippines, Taiwan, and Vietnam.[5]

The Political, Economic, and Military Implications of Joint Exploration and Production in the South China Sea

The PRINCE scores indicate that under a cooperative scenario, China will not engage in joint oil exploration in the Paracel and Spratly Islands. If

China and the nations of Southeast Asia (Malaysia, the Philippines, Taiwan and Vietnam) work together in the South China Sea in joint oil exploration and production projects, there would be no need for India, Japan, or the United States to become involved. Not only would such a scenario avert war but shipping lanes would be far safer. In addition, China may be able to make up for lost oil energy imports imposed by this energy future. Overall, joint oil exploration and production in the South China Sea would not result in a political, economic, or military competition for oil energy resources between China, India, and Japan.

GULF OF MEXICO IN A COOPERATIVE SURPLUS FUTURE

The PRINCE model predicts that China will jointly prospect for oil in the Gulf of Mexico with U.S. oil companies by 79 percent.[6]

The Political, Economic, and Military Implications of Joint Exploration and Production in the Gulf of Mexico with Independent U.S. Oil Companies

Although it is impractical for China to explore for and produce oil beyond its shores, in this scenario there is a 79 percent chance that Beijing will engage in joint exploration and extraction of crude oil with independent U.S. oil companies in the Gulf of Mexico. Chinese oil companies such as the China National Petroleum Corporation (CNPC) and the China National Chemical Corporation (Sinopec) and independent oil companies such as ExxonMobil and ChevronTexaco will work as business partners to explore, discover, extract, and refine crude oil in the region. As a precondition for this arrangement, the U.S. Congress must approve any joint business agreements between the Chinese oil companies and independent oil companies since oil is a strategic commodity and a national security issue. Trust-building measures may have to be undertaken first however, which is desired by China, but may not be favored by factions in the U.S. Congress. Should the U.S. Congress permit joint oil exploration operations in its backyard, the potential is that greater political, economic, and military bilateral relations could develop between the two countries.

SUMMARY

The emergence of the surplus scenarios is likely to be a fairly protracted process since the conversion to alternative fuels is likely to be uneven over time. In any event, the widespread emergence of alternative fuels/technologies within the next one hundred years might be ushered in by an increase in domestic violence in some of the rentier oil producers if it is accompanied by a drop in oil revenues which these governments dispense as a form of patronage. Regardless, China, India, and Japan will not be competing for scarce energy resources making it less likely for a change in the international status quo. Global issues that are unrelated to energy security are unlikely to reduce the latent political tensions between China, India, and Japan.

NOTES

1. Prince and Panda website, "PRINCE_SCORES.pdf," accessed October 17, 2013, http://www.princeandpanda.com.
2. Prince and Panda website, "PRINCE_SCORES.pdf," accessed October 17, 2013, http://www.princeandpanda.com.
3. Prince and Panda website, "PRINCE_SCORES.pdf," accessed October 17, 2013, http://www.princeandpanda.com.
4. Prince and Panda website, "PRINCE_SCORES.pdf," accessed October 17, 2013, http://www.princeandpanda.com.
5. Prince and Panda website, "PRINCE_SCORES.pdf," accessed October 17, 2013, http://www.princeandpanda.com.
6. Prince and Panda website, "PRINCE_SCORES.pdf," accessed October 17, 2013, http://www.princeandpanda.com.

11

Conclusions

This book examined how China's oil energy needs will likely shape future Sino-Indian and Sino-Japanese relations under conditions of dependency and non-dependency. Scenario analysis was used to study China's oil energy needs because it provided the best way to establish a quantitative causality between China's alternative oil energy futures or scenarios and their strategic implications for India and Japan. Three plausible alternative future oil energy futures were developed and analyzed, determining their associated events and strategic political, economic and military implications for India and Japan.

Using the PRINCE model of risk analysis forecasting, it was determined empirically that *the Competitive Dependence and Competitive Surplus energy futures have some probability of occurring because the average PRINCE score (of all the independent events and issues) for each scenario was greater than 60 percent.* If a scenario had an average PRINCE score between 40 to 60 percent, then it was uncertain as to whether a scenario will occur. Moreover, if a scenario had an average PRINCE score (of all the independents events and issues) of less than 40 percent, then a scenario would not occur. *It is uncertain as to whether the Cooperative Surplus energy future would occur because of a PRINCE score falling between 40 percent and 60 percent.* This information is useful because it tells us that China will not necessarily follow a path of oil energy dependence according to the status quo. The scenarios are illustrated graphically in figure 11.1 on a continuum based on their probability of support and likelihood of occurrence and in table 11.1. Note that the *Cooperative Dependency energy future* was not considered because it is an unrealistic scenario. See pages 10 and 11 of chapter 1 for a discussion on why this oil energy future is not feasible.

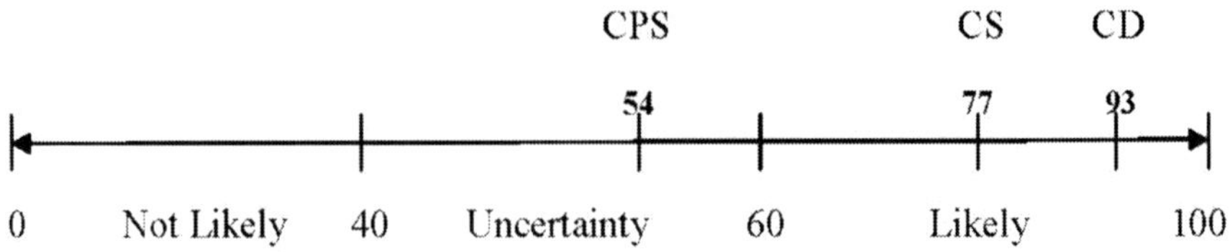

Competitive Dependency (CD)
Competitive Surplus (CS)
Cooperative Surplus (CPS)

Figure 11.1 Probability of Occurrence for Three Chinese Oil Energy Futures

The competitive dependency oil energy future has the highest probability of support average, that is, *93 percent*, based on the average PRINCE scores of seventy-one independent events or issues followed by the competitive surplus (non-dependency) future with probability of support average of *77 percent*, based on the calibrated average PRINCE scores of nine independent events or issues; and the cooperative surplus (non-dependency) future with a probability of support average of *54 percent*, based on the calibrated average PRINCE scores of nine independent events or issues. Even though the competitive dependency energy future has the highest probability of support average of all three alternate energy futures and appears to be the most likely energy future, the competitive surplus energy future is also probable. See table 11.1.

Table 11.1 Chinese Oil Energy Futures or Scenarios

Oil Energy Future/Scenario #1: *Competitive Dependency* (Oil Dependency) *Highest Probability of Support (93%)*	**Oil Energy Future/Scenario #2:** *Competitive Surplus* (Oil Non-Dependency)
Oil Energy Future/Scenario #4: *Cooperative Dependency* (Oil Dependency) *This future/scenario was not considered.*	**Oil Energy Future/Scenario #3:** *Cooperative Surplus* (Oil Non-Dependency)

This research shows that China, India, and Japan will most likely compete for scarce oil energy resources under oil dependency conditions and are less likely to use fuel-saving technologies competitively or cooperatively under oil non-dependency conditions. Obviously, competition between China, India, and Japan for scarce oil energy resources will also generate strategic consequences for poor but resource-rich producer countries which will still be afflicted with the "resource curse." What is most troubling is that China could find itself making political concessions such as arms-for-oil deals. China's energy insecurity could fuel global insecurity if it were to exchange weapons of mass destruction technologies for reliable oil supplies as in the case of Iran.[1]

As previously stated, through the adroit use of soft power, China's strategic interactions with oil supplier states of the non-free and partly-free regions of the world potentially undermines India and Japan's economic efforts to obtain reliable crude oil and petroleum sources. Cases in point are Angola, Nigeria, and Sudan.

Moreover, unless China can find a way to reduce its oil dependence, its hard power and strategic military engagement will grow to ensure its energy security. China has been accelerating its own submarine development and acquiring more Russian submarines capable of operating long distances. China has also been quietly engaged in a strategy of military and economic cooperation with Pakistan, Myanmar, Bangladesh, Sri Lanka, and the Maldives in the Indian Ocean region in order to build and/or have access to ports and airbases to support not only a future tanker fleet, but also a blue-water navy capable of projecting power not only in the Indian Ocean region but in the Persian Gulf.

Chinese warships are currently patrolling the Bab-el-Mandeb in support of international antipiracy operations. Beyond the Indian Ocean region, China established an intelligence collection facility in Cuba and has been offered the use of a former U.S. airbase in Ecuador. China is also engaged in a "weapons for oil exchange" with Iran. It is Iran's largest conventional weapons supplier and its greatest supporter of nuclear technology development. China coordinates its overseas military operations in support of the larger, foreign, diplomatic, economic, and security agenda established by its leadership. Although there are few conditions for a military confrontation over oil energy resources, it is not inconceivable that the Chinese Navy may be able to project sea power and establish sea control well beyond its shores including the sea lines of communication (SLOCs) to ensure its energy security in the near future.

China's rising dependence on imported oil, adroit soft power, economic prowess, strategic engagement of the world as an alternative model of economic development, military modernization, and Mahanian naval desirability in tandem with rapid economic growth as the fastest growing economy in the world in thirty years with an average GDP growth rate above 10 percent can only mean that it will alter the global status quo and become the dominant actor in world affairs.[2] India and Japan will be less influential politically, economically, and militarily because China is skillfully harnessing and strategically exercising the elements of national power (diplomatic, informational, military, and economic) to acquire scarce oil energy resources in the Near East, Western Hemisphere, and Sub-Saharan Africa and is becoming a great power as a consequence.

How India and Japan will react when this change occurs is uncertain. What is certain is that China will be the dominant power in the world unless it can find a way to be less dependent on foreign oil. One possible way is through alternative fuels/technologies. However, the emergence of alternative fuels/technologies future is likely to be a fairly protracted process since the conversion to alternative fuels is likely to be uneven over time. Regardless, the widespread emergence of alternative fuels/technologies might be ushered in by an increase in domestic violence in some of the rentier oil producers if it is accompanied by a drop in oil revenues which these governments dispense as a form of patronage.

The question that emerges from this exhaustive research on China's energy insecurity is what kind of power will it be if it alters the status quo? Will it be a responsible stakeholder of the international system or will it contribute to global insecurity? These are questions that policymakers and government officials in China, India, and Japan must ponder with respect to China's dependence on foreign oil.

INDIAN AND JAPANESE POLICY RESPONSES FOR THE COMPETITIVE DEPENDENCY SCENARIO

The competitive dependency scenario will have the greatest political, economic, and military implications for India and Japan if this scenario should materialize. Therefore, India and Japan should act accordingly by implementing public policies that are consistent with its national security interests. However, any policy response (initiative) should be judicious in nature, cautious in approach, and should not diminish or damage diplomatic ties, economic, and trade relations or global trade. Otherwise, the consequences will be grave for global political and economic stability as well as bilateral relations.

In conjunction with public policy, India and Japan should adopt a sophisticated strategy that utilizes the full array of diplomatic, information, economic, and military tools available in a reasonably coordinated fashion to influence China's energy outcomes.[3] The goals of such a strategy should be:[4]

- To build the mutual trust necessary for effective cooperation;
- To increase China's willingness to work with India and Japan counterparts to develop and deploy capabilities to address China's oil energy problems;

- To enhance incentives and opportunities for China to follow a desired energy strategy; and
- To help China develop specific technical and policy capabilities that will better enable it to achieve desirable energy outcomes.

Foremost to any policy with respect to China is trust. According to Lieberthal and Herberg, "trust is both fundamental and nettlesome."[5] China should trust India and Japan enough to allow effective cooperation on any energy matters. Trust becomes important in reducing China's acute security concerns regarding the sourcing and transporting of its imported oil energy.[6] Trust between China, India, and Japan should be established to provide for more effective cooperation on oil energy matters.

RECOMMENDATIONS FOR TRILATERAL INITIATIVES IN ENERGY COOPERATION

A high level of engagement is critical because no sector is more intimately tied to the fate of the global economy than energy, particularly oil energy. To this extent, India and Japan should:

- Convene bilateral dialogues on energy cooperation to build trust;
- Promote cooperative dialogues on oil-saving technologies (particularly transportation technology);
- Encourage China to take on a greater role within global economic, trade, and financial institutions and mechanisms, including the Organization for Economic Co-operation and Development Export Credit Arrangement, International Energy Agency (IEA), World Bank and G-8 process; and
- Encourage greater military-to-military ties with China, especially regarding the observation and participation in peacetime military training, exercises and operations related to the area of energy security.

Each of the initiatives listed above are self-explanatory but it is helpful to explain the role and efficacy of each initiative with respect to bilateral energy cooperation.

Trilateral Dialogues (Phase One)

Trilateral dialogues would be an important first step to build trust in energy cooperation. Dialogues would be non-statutory in nature, conducted formally or informally, and based on shared concerns on energy issues affecting both countries such as demand, consumption, and security.

Cooperative Dialogues on Oil-Saving Technologies (Phase Two)

In addition to promoting trilateral dialogues, India and Japan can also promote specific dialogues on oil-saving technologies as the next step in energy cooperation. Dialogues should promote the development and adoption of plug-in hybrid engines, the sustainable production of biofuels and fuel efficiency improvements with the shared goal of reducing demand growth of fossil fuels.

Trilateral Military Cooperation

Military cooperation especially naval cooperation in the area of energy security would help to build trust between China, India, and Japan since China does not have the ability to protect the SLOCs militarily in the event of a supply disruption caused by piracy, terrorism or naval blockade. Typically, the security of the SLOCs falls upon the United States Navy and its allies in the form of Combined Task Forces (CTFs) operating in the Gulf of Aden, Gulf of Oman, the Arabian Sea, the Indian Ocean, off the Somali Coast and the Arabian Gulf. If India and Japan could encourage China to operate jointly with one or more of the Combined Task Forces rather than independently, then trust could be built between them and alleviate some concerns that China may have with respect to potential supply disruptions by U.S. naval forces. Further, China's naval participation would help to reduce U.S. expenditures to defend oil supplies from the Arabian Gulf.

OECD Export Credit Arrangement

This is a special purpose regime composed of nine participants (Australia, Canada, the European Community, Japan, South Korea, Norway, New Zealand, Switzerland, and the United States) with two basic objectives: to control the predatory use of export finance to shift rents from one supplier state to another; and to control the unintentional transfers of wealth from supplier to buyer states.[7] It sets and enforces terms and conditions on official export credits and foreign aid.[8] Further, it plays an important role in the multilateral trading system to provide a level playing field so that both OECD and non-OECD exporters compete on the price and quality of their goods and services, not on the support they receive from their governments.[9]

The success of the Export Credit Arrangement derives from its participants' aims to eliminate trade distortion and level the playing field, driven by economic, competitive, and developmental motivations and the need to reach practical and pragmatic consensus.[10] Although China is not a member in the Export Credit Arrangement, membership would help it become a responsible stakeholder in the international system by conforming to global arrangements governing state-backed finance because state-backed finance can distort energy markets.[11] Thus, China's use of predatory financing would be diminished and would not undermine an open and competitive world oil market.

RECOMMENDATIONS FOR MULTILATERAL INITIATIVES IN ENERGY COOPERATION

Multilateral initiatives offer a very important basis for building trust in the energy world in light of the many problems in China's relations with India and Japan. Therefore, it behooves India and Japan to consider and promote the following initiatives because each has the potential to produce positive outcomes in energy cooperation and to reduce Chinese mistrust of long-term Indian and Japanese intentions in the energy world. To this extent, India and Japan should:

- Convene multilateral dialogues on energy cooperation to build trust;
- Encourage international partnerships between national oil companies and independent oil companies that demonstrate the win-win potential of international cooperation in oil investment;
- Broaden diplomatic dialogue inside regions where there are significant oil energy investments to identify overlapping interests, such as stable investment conditions, and possible areas for cooperation, including foreign assistance; and
- Persuade global economic, trade, and financial institutions such as the G-8 and International Energy Agency (IEA) to invite China as a member to show how the international system can directly benefit China's energy security.

Each of initiatives listed above are self-explanatory but it is helpful to explain them further in terms of their role and efficacy in terms of multilateral energy cooperation.

Multilateral Dialogues

Multilateral dialogues conducted in international forums like the United Nations and the World Trade Organization would be an important initial step to building trust in energy cooperation between China, India, and Japan. The advantage to conducting multilateral dialogues in international forums like the United Nations and the World Trade Organization as opposed to the G-8 and the International Energy Agency is that China is already a member of the former two organizations. Membership to the latter two organizations requires China to become an advanced industrialized economy in the case of the G-8 or a member of the Organization for Economic Co-operation and Development (OECD) in the case of the International Energy Agency. To qualify for membership in the OECD, China must have a per capita GDP of at least $3,000 and meet certain standards of human rights. Therefore, "any effort to relax existing conditions to accommodate Chinese membership would prove to be too divisive among current IEA members."[12]

International Partnerships

Encouraging international partnerships between national oil companies and international oil companies not only demonstrates the win-win potential of international cooperation but also helps to solidify the rules of the game with respect to oil energy investments.

Regional Diplomatic Dialogues

Broadening diplomatic dialogues in oil rich regions to identify overlapping interests, such as stable investment conditions, and possible areas for cooperation, including foreign assistance with respect to oil energy investments can bring about deconfliction as well as building trust and cooperation between China, India, Japan and oil energy producers.

The G-8

The G-8 is an informal forum representing the most advanced industrialized democracies that meets annually to address major economic issues of common concern. The G-8 would benefit from revising its institutional mission by becoming a forum that includes not only the advanced industrial democratic states but also the most consequential countries in the world to address the most pressing economic issues such as energy.[13] Japan can play an important role by persuading the G-8 to invite China as well as encouraging China to participate in the G-8 and thereby become a major global stakeholder. It does not make sense to exclude China from an international regime like the G-8 if such exclusion encourages China to a free ride on the benefits of that regime without bearing any serious responsibility for its effective contributions or enforcement.[14]

The IEA

The International Energy Agency is a Paris-based intergovernmental organization established under the OECD in the wake of the 1974 oil crisis.

It was initially focused on formulating multilateral measures that would reduce the vulnerability of its member countries to future oil supply disruptions but has since broadened its focus on energy policy in the following areas: energy security, economic development, and environmental protection (International Energy Agency).

Although China has participated in IEA workshops and related activities, China is not a member of the IEA because it is not a member of OECD. If India and Japan could persuade the IEA to make China an actual IEA member with full participation in all activities and full assumption of responsibilities for the management of supply disruptions, then India and Japan would be sending a strong signal to China that India and Japan is serious about treating China as an important player in the international energy regime.[15]

RECOMMENDATIONS FOR REGIONAL INITIATIVES

In this energy future scenario, regional initiatives are proposed for the Asia-Pacific, Central Eurasia, the Near East (Middle East and North Africa), Sub-Saharan Africa (Africa), Western Europe and the Western Hemisphere. The initiatives will vary in size and scope in light of overall Chinese, Indian, and Japanese strategic engagement and available oil energy reserves in each region.

Asia (includes Oceania, Pakistan, Russia and the South China Sea)

Asia is one of the least energy rich regions of the world with approximately 3 percent of the world's proven oil reserves. Nevertheless, Asia is vital to China's strategic interests because most of China's imported crude oil and petroleum is transported by ships transiting through the Strait of Malacca and the South China Sea. Given the lack of trust China, India, and Japan have with respect to energy security and the potential for accidental military escalation in the East China Sea and South China Sea, it would behoove India and Japan to undertake key multilateral initiatives in the region not only to build trust but to minimize the potential for accidental war. Such initiatives would include:

- Involve China in joint naval efforts to patrol the Strait of Malacca to contribute to the stability of sea lines of communication;
- Create patterns of cooperation with China on regional security issues such as energy security so China's growing role is in the direction of cooperation rather than conflict;

- Ensure regular, senior-level contact with Chinese and Asian leaders to anticipate and preempt potential areas of conflict and disagreement especially with respect to the East China Sea and the South China Sea; and
- Empower regional institutions, such as the ASEAN Regional Forum (ARF), through forms of multilateral cooperation to address energy security challenges.

Each of initiatives listed above are self-explanatory but it is helpful to explain them further in terms of their benefits to Indian and Japanese energy security.

Involve China in joint naval efforts to patrol the Strait of Malacca to contribute to the stability of sea lines of communication

Naval cooperation in the area of energy security would help to build trust between China, India, and Japan since China does not have the ability to protect the SLOCs militarily in the event of a supply disruption caused by piracy, terrorism, or naval blockade. Typically, the security of the SLOCs falls upon the United States Navy and its regional allies. If India and Japan could encourage China to operate jointly, then trust could be built between them and alleviate some concerns that China may have with respect to potential supply disruptions by U.S. naval forces.

Create patterns of cooperation with China on regional security issues such as energy security so China's growing role is in the direction of cooperation rather than conflict

Chinese as well as Indian and Japanese oil energy demand and consumption will increase over the next twenty years. Further, China's role on the international stage is expected to rise in conjunction with increased energy demand and consumption which may or may not have adverse consequences for bilateral relations. It would be beneficial to India and Japan to find novel ways, both formally and informally, to indulge China to participate in confidence-building measures (CBMs) to reduce tensions and the specter of an accidental war in the region.

Ensure regular, senior-level contact with Chinese and Asian leaders to anticipate and preempt potential areas of conflict and disagreement especially with respect to the East China Sea and the South China Sea

Regular, senior-level contact with Chinese and Asian leaders, especially military leaders in the form of military-to-military relations, is vital to deconfliction especially since the potential for accidental war is great.

Further, track-two dialogues (informal relations) can influence other actors in the security and defense establishments, such as naval officers through military-to-military maritime dialogues.[16] In fact, track-two dialogues involving maritime confidence-building proved less contentious than other regional issues and appeared to be a productive area for cooperative security practices in South Asia.[17]

Empower regional institutions, such as the ASEAN Regional Forum (ARF), through forms of multilateral cooperation to address energy security challenges

The ASEAN Regional Forum (ARF) has been regarded as a better model for security cooperation in non-Western regions than the European experience of highly institutionalized regional cooperation because of its emphasis on personal contacts, informality, and consensus-building rather than formal institutionalized decision-making.[18] Consisting of twenty-seven participants, ARF objectives are to foster dialogue and consultation, and promote confidence-building and preventive diplomacy in the region. If India and Japan could empower ARF to address pressing energy security challenges in Southeast Asia, it may help to reduce the possibility of accidental war in the South China Sea and possibly the East China Sea.

Central Eurasia

Central Eurasia is not particularly energy rich with approximately 2.9 percent of the world's proven oil reserves. Nevertheless, Central Eurasia is important to China in terms of energy security because China has unfettered access to the region's crude oil reserves without interference from India and Japan. Given the lack of trust between them with respect to energy security, it would behoove India and Japan to undertake key multilateral initiatives to build bilateral trust in the region. Such initiatives would include:

- Create patterns of cooperation with China on regional security issues such as energy security so China's growing role is in the direction of cooperation rather than conflict, and
- Empower regional institutions, such as the Shanghai Cooperation Organization (SCO), through forms of multilateral cooperation to address energy security challenges.

Each of the initiatives listed above are self-explanatory but it is helpful to explain them further in terms of their benefits to Indian and Japanese energy security.

Create patterns of cooperation with China on regional security issues such as energy security so China's growing role is in the direction of cooperation rather than conflict

China's role on the international stage is expected to rise in conjunction with increased energy demand and consumption which may or may not have adverse consequences for bilateral relations. It would be beneficial to India and Japan to find novel ways, both formally and informally, to indulge China to participate in confidence-building measures (CBMs) to reduce tensions and build trust.

Empower regional institutions, such as the Shanghai Cooperation Organization (SCO), through forms of multilateral cooperation to address energy security challenges

Although the Shanghai Cooperation Organization is primarily focused on its member nations' Central Asian security-related concerns like terrorism, separatism, and extremism, expanded dialogue regarding the potential energy security challenges facing the region would help to improve relations not only between members but also build trust. India and Japan could play an important role in this respect even though they are not SCO members.

The Near East (Middle East and North Africa)

The Near East is the most energy rich region with approximately 54 percent of the world's proven oil reserves. Further, the Near East produces almost 50 percent of China's crude oil and petroleum. Because the region produces almost 50 percent of China's oil, it is critical to China's energy security especially since the oil is transported by ships along the sea lines of communication from the Persian Gulf to the South China Sea and is subject to potential supply disruptions by pirates, terrorists, and the United States Navy. Given the lack of trust between China, India, and Japan with respect to energy security, China's growing influence in the region with major oil producers like Iran and Saudi Arabia and the potential for accidental war between the United States and Iran, it would behoove India and Japan to undertake key multilateral initiatives in the region not only to build trust between them but to minimize the potential for accidental war. Such initiatives would include:

- Involve China in joint naval efforts to patrol the Arabian Gulf to contribute to the stability of sea lines of communication;

- Encourage international partnerships between national oil companies and independent oil companies that demonstrate the win-win potential of international cooperation in oil investment in the Near East;
- Create patterns of cooperation with China on regional security issues such as energy security so China's growing role is in the direction of cooperation rather than conflict; and
- Ensure regular, senior-level contact with Chinese and Near East leaders to anticipate and preempt potential areas of conflict and disagreement especially with respect to the possibility of accidental military conflict in the Arabian Gulf.

Each of the initiatives listed above are self-explanatory but it is helpful to explain them further in terms of their benefits to Indian and Japanese energy security.

Involve China in joint naval efforts to patrol the Arabian Gulf to contribute to the stability of sea lines of communication

Joint naval cooperation in the Arabian Gulf would help to reduce tensions and build trust especially since the United States and its allies patrol the region. China's naval participation would help to reduce U.S. expenditures to defend oil supplies from the Arabian Gulf.

Encourage international partnerships between national oil companies and independent oil companies that demonstrate the win-win potential of international cooperation in oil investment in the Near East

Encouraging international partnerships between national oil companies and international oil companies in the Near East not only demonstrates the win-win potential of international cooperation but also helps to solidify the rules of the game with respect to oil energy investments.

Create patterns of cooperation with China on regional security issues such as energy security so China's growing role is in the direction of cooperation rather than conflict

China's role in the Near East is expected to rise in conjunction with increased energy demand and consumption which may or may not have adverse consequences for Sino-Indian or Sino-Japanese relations. It would be beneficial to India and Japan to find novel ways, both formally and informally, to indulge China to participate in confidence-building measures to reduce tensions and build trust especially since the Near East produces approximately 50 percent of China's oil.

Ensure regular, senior-level contact with Chinese and Near East leaders to anticipate and preempt potential areas of conflict and disagreement especially with respect to the possibility of accidental military conflict in the Arabian Gulf

Regular, senior-level contact with Chinese and Near East leaders, especially military leaders in the form of military-to-military relations, is vital to deconfliction especially since the potential for accidental war is great in the Arabian Gulf.

Sub-Saharan Africa (Africa)

Africa is one of the least energy rich regions of the world with approximately 4 percent of the world's proven oil reserves. However, the region produces almost 30 percent of China's crude oil and petroleum. Because the region produces almost 30 percent of China's oil, it is critical to China's energy security especially since oil is transported by ships along the sea lines of communication from key oil producing states like Angola, Nigeria, and the Sudan and is subject to potential supply disruptions by pirates, terrorists, and the United States Navy. Given the lack of trust between China, India, and Japan with respect to energy security, China's growing influence with major oil producers like Angola, Nigeria, and Sudan and the potential for supply disruptions, it would behoove India and Japan to undertake key multilateral initiatives in the region to build trust between them. Such initiatives would include:

- Expand diplomatic presence in Africa;
- Encourage and support India and Japan (especially IOCs) to invest more in Africa;
- Involve China in joint naval efforts to patrol coastal Africa in addition to the Horn of Africa to contribute to the stability of sea lines of communication;
- Encourage international partnerships between national oil companies and independent oil companies that demonstrate the win-win potential of international cooperation in oil investment in Africa;
- Create patterns of cooperation with China on regional security issues such as energy security so China's growing role is in the direction of cooperation rather than conflict; and
- Ensure regular, senior-level contact with Chinese and African leaders to anticipate and preempt potential areas of conflict and disagreement.

Each of the initiatives listed above are self-explanatory but it is helpful to explain them further in terms of their benefits to Indian and Japanese energy security.

Expand diplomatic presence in Africa

Compared to other regions of the world, Africa receives the least amount of Indian and Japanese political and diplomatic attention. India and Japan need to reevaluate how it is viewed in Africa and focus more on its diplomatic imperatives. They need to expand their diplomatic presence in Africa and devote more attention to its diplomatic relations with individual African states.

Encourage and support Indian and Japanese businesses (especially IOCs) to invest more in Africa

Indian and Japanese businesses can be important sources of revenue to fund needed infrastructure projects in some of the poorest and underdeveloped areas of Africa in addition to public sector funding. Funding not only helps the impoverished and malnourished people of Africa but it also helps to built political goodwill between nations. If India and Japan could encourage more businesses, especially IOCs, to invest more in Africa, it would improve their relations with Africa.

Involve China in joint naval efforts to patrol coastal Africa in addition to the Horn of Africa to contribute to the stability of sea lines of communication

Joint naval cooperation to patrol coastal Africa along the SLOCs would reduce tension and build trust between China, India, and Japan. However, China operates somewhat independently rather than jointly around the Horn of Africa for the prosecution of international pirates.

Encourage international partnerships between national oil companies and independent oil companies that demonstrate the win-win potential of international cooperation in oil investment in Africa

Encouraging international partnerships between national oil companies and international oil companies in Africa not only demonstrates the win-win potential of international cooperation but also helps to solidify the rules of the game with respect to oil energy investments.

Create patterns of cooperation with China on regional security issues such as energy security so China's growing role is in the direction of cooperation rather than conflict

China's role in Africa is rising in conjunction with increased energy demand and consumption which may or may not have adverse consequences for bilateral relations. It would be beneficial to India and Japan to find novel ways, both formally and informally, to indulge China to participate in confidence-building measures to reduce tensions and build trust especially since Africa produces approximately 30 percent of China's oil.

Ensure regular, senior-level contact with Chinese and African leaders to anticipate and preempt potential areas of conflict and disagreement

Regular, senior-level contact with Chinese and African leaders, especially military leaders in the form of military-to-military relations, serve as important mechanisms for building trust and reducing tensions with respect to energy security matters especially since many oil exploration and production activities are in politically unstable regions like Nigeria and the Sudan.

Western Europe

Western Europe is the least energy rich region of the world with approximately 1 percent of the world's proven oil reserves. Nevertheless, Western Europe is important to China's strategic interests. Given the lack of trust between China, India, and Japan with respect to energy security, it would behoove India and Japan to undertake key multilateral initiatives in the region to build trust between them. Such initiatives would include:

- Ensure regular, senior-level contact with Chinese and European leaders to anticipate and preempt potential areas of conflict and disagreement; and
- Promote and organize multilateral cooperative dialogues and exchanges on oil-saving technologies (particularly transportation technology).

Each of the initiatives listed above are self-explanatory but it is helpful to explain them further in terms of their benefits to Indian and Japanese energy security.

Ensure regular, senior-level contact with Chinese and European leaders to anticipate and preempt potential areas of conflict and disagreement

Regular, senior-level contact with Chinese and European leaders, especially military leaders in the form of military-to-military relations, serve as important mechanisms for building trust and reducing tensions with respect to energy security matters.

Promote and organize multilateral cooperative dialogues and exchanges on oil-saving technologies (particularly transportation technology)

Indian- and Japanese-led promotional and organizational efforts in multilateral cooperative dialogues and exchanges on oil-saving technologies in addition to unilateral domestic efforts (see above) would help to create a perception that India and Japan are serious about alternative fuel technology not only in their countries but also in other developed regions on the world like Western Europe.

Western Hemisphere

The Western Hemisphere is the most energy rich region in the world after the Near East with approximately 31 percent of the world's proven oil reserves. Although the Western Hemisphere produces less than 7 percent of China's oil, it is important to China's energy security and strategic interests. Given the lack of trust between China, India, and Japan with respect to energy security and China's growing influence in the region with major oil producers like Venezuela, it would behoove India and Japan to undertake key multilateral initiatives in the region to build trust between them. Such initiatives would include:

- Encourage international partnerships between national oil companies and independent oil companies that demonstrate the win-win potential of international cooperation in oil investment in Latin America;
- Create patterns of cooperation with China with respect to energy trading so China's growing role is in the direction of cooperation rather than conflict; and
- Ensure regular, senior-level contact with Chinese and Latin American leaders to anticipate and preempt potential areas of conflict and disagreement.

Each of the initiatives listed above are self-explanatory but it is helpful to explain them further in terms of their benefits to Indian and Japanese energy security.

Encourage international partnerships between national oil companies and independent oil companies that demonstrate the win-win potential of international cooperation in oil investment in Latin America

Encouraging international partnerships between national oil companies and international oil companies in Latin America not only demonstrates the win-win potential of international cooperation but also helps to solidify the rules of the game with respect to oil energy investments.

Create patterns of cooperation with China with respect to energy trading so China's growing role is in the direction of cooperation rather than conflict

China's role in Latin America is rising in conjunction with increased energy demand and consumption which may or may not have adverse consequences for Sino-Indian and Sino-Japanese relations. It would be beneficial to India and Japan to find novel ways, both formally and informally, to indulge China to participate in confidence-building measures to reduce tensions and build trust.

Ensure regular, senior-level contact with Chinese and Latin American leaders to anticipate and preempt potential areas of conflict and disagreement

Regular, senior-level contact with Chinese and Latin American leaders, serve as important mechanisms for building trust and reducing tensions with respect to energy security matters especially since China is increasing its political and economic presence in Latin America.

INDIAN AND JAPANESE RESPONSES FOR THE SURPLUS SCENARIOS

For the most part, the surplus scenarios will have positive political, economic, and military implications for India and Japan if this scenario should come to fruition. India and Japan act accordingly by implementing trilateral and multilateral policies that are consistent with their national security interests. However, any policy response (initiative) should be judicious in nature, cautious in approach, and should not diminish or damage diplomatic ties, economic and trade relations, or global trade. Otherwise, the consequences will be grave for global political and economic stability as well as Sino-Indian and Sino-Japanese relations.

In conjunction with public policy, India and Japan should adopt a sophisticated strategy that utilizes the full array of diplomatic, information, economic, and military tools available in a reasonably coordinated fashion to influence China's energy outcomes. The goals of such a strategy should be to fundamentally build the mutual trust necessary for effective cooperation and to help China develop specific technical and policy capabilities that will better enable it to achieve desirable energy outcomes.[19]

RECOMMENDATIONS FOR TRILATERAL INITIATIVES IN ENERGY COOPERATION

A high level of engagement is critical because no sector is more intimately tied to the fate of the global economy than energy, particularly oil energy.

To this extent, India and Japan should:

- *Phase One:* Convene trilateral dialogues on energy cooperation to build trust.

  Trilateral dialogues would be an important first step to build trust in energy cooperation. Dialogues would be non-statutory in nature, conducted formally or informally, and based on shared concerns on energy issues affecting both countries such as demand, consumption, and security.
- *Phase Two*: Promote cooperative dialogues on oil-saving technologies (particularly transportation technology).

  In addition to promoting trilateral dialogues, India and Japan can also promote specific dialogues on advanced oil-saving technologies as the next step in energy cooperation. Dialogues should promote the continued development and potential use of plug-in hybrid engines in addition to hybrid engines, the sustainable production of biofuels and fuel efficiency improvements with the shared goal of reducing demand growth of fossil fuels.
- *Phase Three*: Encourage greater military-to-military ties with China, especially regarding the observation and participation in peacetime military training, exercises and operations related to the area of energy security.

  Military cooperation especially naval cooperation in the area of energy security would help to build trust between China, India, and Japan since the potential for an accidental war is great in the East China Sea and South China Sea.

RECOMMENDATIONS FOR MULTILATERAL INITIATIVES IN ENERGY COOPERATION

Multilateral initiatives offer an important basis for building trust in the energy world in light of the many problems in trilateral relations.

Therefore, it behooves India and Japan to consider and promote the following initiatives because each has the potential to produce positive outcomes in energy cooperation and to reduce Chinese mistrust of long-term Indian and Japanese intentions in the energy world. To this extent, India and Japan should:

- *Phase One*: Convene multilateral dialogues on energy cooperation to build trust.

  Multilateral dialogues conducted in international forums like the United Nations and the World Trade Organization would be an important initial step to building trust in energy cooperation between China, India, and Japan. The advantage to conducting multilateral dialogues in international forums like the United Nations and the World Trade Organization as opposed to the G-8 and the International Energy Agency is that China is already a member of the former two organizations. Membership to the latter two organizations requires China to become an advanced industrialized economy in the case of the G-8 or a member of the Organization for Economic Cooperation and Development (OECD) in the case of the International Energy Agency. To qualify for membership in the OECD, China must have a per capita GDP of at least $3,000 and meet certain standards of human rights. Therefore, "any effort to relax existing conditions to accommodate Chinese membership would prove to be too divisive among current IEA members."[20]
- *Phase Two*: Encourage international partnerships between national oil companies and independent oil companies that demonstrate the win-win potential of international cooperation in oil investment (*Competitive Surplus scenario*).

  Encouraging international partnerships between national oil companies and international oil companies for projects in mainland China and China's territorial waters not only demonstrate the win-win potential of international cooperation but also helps to solidify the rules of the game with respect to oil energy investments.
- *Phase Three*: Broaden diplomatic dialogue inside regions where there are significant oil energy investments to identify overlapping interests, such as stable investment conditions, and possible areas for cooperation, including foreign assistance (*Competitive Surplus Scenario*).

  Broadening diplomatic dialogues in oil rich regions of East Asia and Southeast Asia to identify overlapping interests, such as stable investment conditions, and possible areas for cooperation, including foreign assistance with respect to oil energy investments can bring about deconfliction as well as building trust and cooperation between China, India, Japan, and oil energy producers.

RECOMMENDATIONS FOR REGIONAL INITIATIVES IN ENERGY COOPERATION

In Surplus scenarios, regional initiatives are proposed for Asia-Pacific. The initiatives will vary in size and scope in light of overall Chinese, Indian, and Japanese strategic engagement and available oil energy reserves in each region.

Asia-Pacific

Asia-Pacific is one of the least energy rich regions of the world with approximately 3 percent of the world's proven oil reserves. Nevertheless, Asia is vital to China's strategic interests. Given the potential for a kinetic military exchange in the East China Sea and South China Sea, it would behoove India and Japan to undertake key multilateral initiatives in this region not only to build trust between them but to minimize the potential for accidental war. Such initiatives would include:

- Create patterns of cooperation with China on regional security issues such as energy security so China's growing role is in the direction of cooperation rather than conflict since oil energy demand and consumption will increase over the next twenty years. China's role on the international stage is expected to rise in conjunction with increased energy demand and consumption which may or may not have adverse consequences for Sino-Indian and Sino-Japanese relations. It would be beneficial to India and Japan to find novel ways, both formally and informally, to indulge China to participate in confidence-building measures to reduce tensions and the specter of an accidental war in the region.
- Empower regional institutions, such as the ASEAN Regional Forum (ARF), through forms of multilateral cooperation to address energy security challenges. ARF has been regarded as a better model for security cooperation in non-Western regions than the European experience of highly institutionalized regional cooperation because of its emphasis on personal contacts, informality, and consensus-building.[21] Consisting of twenty-seven participants, ARF objectives are to foster dialogue and consultation, and promote confidence-building and preventive diplomacy in the region.

WHAT CAN CHINA DO?

China has already taken steps to constrain the rise of oil use in the transport sector. As previously discussed, China has raised the level of purchase tax on large cars, raised the fuel economy standards for vehicle manufacturers, increased state funding for research and development in alternative-fuelled vehicles, and ordered that government departments use vehicles with higher fuel efficiency standards and that departmental vehicles not be available for private use by employees.[22]

China could also adopt public policy measures like offering tax credits to citizens to use public transportation and to purchase fuel-efficient vehicles, including alternative-fuelled vehicles. China could also require government workers to carpool to work or use public transportation.

NOTES

1. Erica Strecker Downs, "China's Energy Security" (PhD diss, Princeton University, 2004), 243.
2. "Chinese Economy Slows To Still Sizzling 11.5 percent Growth," *USA Today*, October 25, 2007. *http://usatoday30.usatoday.com/money/world/2007-10-25-china-gdp_N.htm.*
3. Kenneth Lieberthal and Mikkal Herberg, *China's Search for Energy Security: Implications for U.S. Policy* (Seattle: National Bureau of Asian Research), Volume 17, Number 1, April 2006, 28.
4. Lieberthal and Herberg, *China's Search for Energy Security*, 28.
5. Lieberthal and Herberg, *China's Search for Energy Security*, 29.
6. Lieberthal and Herberg, *China's Search for Energy Security*, 28.
7. Peter C. Evans and Erica S. Downs, *Untangling China's Quest for Oil through State-backed Financial Deals* (Washington, DC: The Brookings Institution Press), Policy Brief 154, May 2006, 6.
8. Evans and Downs, *Untangling China's Quest for Oil*, 6.
9. Organization for Economic Co-operation and Development, *The Export Credits Arrangement*, 1978–2008.
10. Organization for Economic Co-operation and Development, *The Export Credits Arrangement*, 1978–2008.
11. Evans and Downs, *Untangling China's Quest for Oil*, 7.
12. Lieberthal and Herberg, *China's Search for Energy Security*, 33.
13. Lieberthal and Herberg, *China's Search for Energy Security*, 32.
14. Lieberthal and Herberg, *China's Search for Energy Security*, 32.
15. Lieberthal and Herberg, *China's Search for Energy Security*, 34.
16. Dalia Dassa Kaye, *Talking to the Enemy: Track-Two Diplomacy in the Middle East and South Asia* (Santa Monica, CA: RAND Corporation, 2007), 93.
17. Kaye, *Talking to the Enemy*, 93.
18. Kaye, *Talking to the Enemy*, 13.
19. Lieberthal and Herberg, *China's Search for Energy Security*, 28.
20. Lieberthal and Herberg, *China's Search for Energy Security*, 33.
21. Kaye, *Talking to the Enemy*, 13.
22. Philip Andrews-Speed and Ronald Dannreuther, *China, Oil and Global Politics* (London: Routledge, 2011), 31.

Appendix A

WORLD'S PROVED RESERVES OF OIL BY REGION (MOST RECENT ESTIMATES)

Region	Country	*EIA* Oil (Billion Barrels)	Oil (Percentage) of World Total
Asia & Oceania	Australia	1.433	0.08
Asia & Oceania	Brunei	1.100	0.06
Asia & Oceania	Burma (Myanmar)	0.050	0.00
Asia & Oceania	China	25.584	1.58
Asia & Oceania	Indonesia	4.030	0.25
Asia & Oceania	Korea, South	0	0.00
Asia & Oceania	Malaysia	4.000	0.24
Asia & Oceania	Mongolia	0	0.00
Asia & Oceania	Philippines	0.139	0.008
Asia & Oceania	Thailand	0.453	0.03
Asia & Oceania	Vietnam	4.400	0.27
	Total	**41.189**	**2.54**
Eurasia	Azerbaijan	7.000	0.43
Eurasia	Kazakhstan	30.000	1.86
Eurasia	Russia	80.000	4.94
Eurasia	Turkmenistan	0.600	0.04
Eurasia	Uzbekistan	0.594	0.04
	Total	**118.194**	**7.30**

Region	Country	*EIA* Oil (Billion Barrels)	Oil (Percentage) of World Total
Middle East & North Africa	Algeria	12.200	0.75
Middle East & North Africa	Egypt	4.400	0.27
Middle East & North Africa	Iran	154.580	9.56
Middle East & North Africa	Iraq	141.350	8.74
Middle East & North Africa	Kuwait	104.000	6.43
Middle East & North Africa	Libya	48.010	2.96
Middle East & North Africa	Morocco	0.00068	0.00
Middle East & North Africa	Oman	5.500	0.34
Middle East & North Africa	Qatar	25.380	1.57
Middle East & North Africa	Saudi Arabia	267.910	16.57
Middle East & North Africa	Syria	2.500	0.15
Middle East & North Africa	Tunisia	0.425	0.03
Middle East & North Africa	United Arab Emirates	97.800	6.04
Middle East & North Africa	Yemen	3.000	0.19
	Total	**867.055**	**53.61**

Region	Country	***Oil and Gas Journal*** **Oil (Billion Barrels)**	Oil (Percentage) of World Total
Sub Saharan Africa	Angola	10.470	0.647
Sub Saharan Africa	Chad	1.500	0.09
Sub Saharan Africa	Congo (Brazzaville)	1.600	0.09
Sub Saharan Africa	Cote d'Ivoire (Ivory Coast)	0.100	0.006
Sub Saharan Africa	Equatorial Guinea	1.100	0.07
Sub Saharan Africa	Gabon	2.000	0.12
Sub Saharan Africa	Guinea	0	0.00
Sub Saharan Africa	Kenya	0	0.00
Sub Saharan Africa	Madagascar	0	0.00
Sub Saharan Africa	Mali	0	0.00
Sub Saharan Africa	Mauritania	0.020	0.001
Sub Saharan Africa	Niger	0	0.00
Sub Saharan Africa	Nigeria	37.200	2.3
Sub Saharan Africa	South Africa	0.015	0.00
Sub Saharan Africa	Sudan	5.000	0.31
	Total	**59.005**	**3.65**

Region	Country	*Oil and Gas Journal* Oil (Billion Barrels)	Oil (Percentage) of World Total
Western Europe	Norway	5.366	0.33
Western Europe	United Kingdom	3.122	0.19
	Total	**8.488**	**.524**
Western Hemisphere	Canada	173.105	10.70
	Mexico	10.264	0.63
	Argentina	2.805	0.17
	Bolivia	0.210	0.01
	Brazil	13.154	0.81
	Colombia	2.200	0.14
	Cuba	0.124	0.007
	Ecuador	8.240	0.51
	Guatemala	0.083	0.005
	Peru	0.579	0.04
	Venezuela	297.570	18.4
	Total	**508.334**	**31.43**
All Regions	**Sub Total**	**1,602.265**	**99.07**
	World Total	**1,617.170**	**100.00**

Source: Energy Information Administration, U.S. Department of Energy, October 5, 2013.

Appendix B

CHINESE FOREIGN OIL ACQUISITIONS AND EXPLORATION AND PRODUCTION PROJECTS

Date	Company	Country/ Region	Description
November 2013	Sinopec	Egypt	Acquired 33% of Apache for $2.9 billion.
February 2013	CNOOC	Gulf of Mexico, Offshore West Africa, and North Sea	Purchased 100% of Nexen, Inc. for $15.1 billion.
December 2011	Sinopec	Canada	Acquired Daylight Energy for $2.1 billion.
October 2011	Sinopec	Indonesia	Acquired 18% stake in the Gendalo-Gehem Project.
November 2010	CNOOC	Argentina	Acquired 60% of Pan American Energy from BP under Bridas which CNOOC has 50% share for $2.47 billion.
November 2010	CNOOC	South Texas, United States	Purchased 33.3% interest in Chesapeake's 600,000 net acres in the Eagle Ford Shale for $2.16 billion.
October 2010	CNOOC	Uganda	Purchased 66% of Tullow Oil's stake in three blocks (1, 2, & 3A) with Total for more than $1 billion.
October 2010	Sinopec	Brazil	Purchased 40% stake of Brazilian subsidiary of Spanish oil company Repsol for $7.1 billion.

Date	Company	Country/ Region	Description
September 2010	CNPC	Russia	The Russia-China Crude Pipeline became operational from Skovorodino in Russia to Mohe at the Sino-Russian border.
August 2010	Sinopec	Kazakhstan	Acquires 50% of Capital Investment Resources Ltd.
May 2010	CIC	Alberta, Canada	Has 45% stake of Canada's Penn West Energy Trust to jointly develop its bitumen assets in the Peace River Region for $800 million.
May 2010	CNPC	Syria	Purchased 35% stake of Shell's wholly owned subsidiary, Syria Petroleum Development BV. (SPD) for a reported $1.2–1.5 billion. SPD owns three production licenses in Syria covering 40 Oil fields with 23 kb/d output in 2009.
May 2010	Sinochem	Brazil	Purchased Statoil's 40% stake in Brazil's Peregrine Oil field for $3.07 billion. Statoil will still retain 60% share and remain as the field operator.
April 2010	Sinopec and Total	Canada	Purchased 9.03% in Canadian oil sands company Syncrude from ConocoPhillips for $4.675 billion. Total is the partner who holds 50%.
March 2010	CNOOC	South America	Purchased 50% stake for $3.1 billion in the Argentinean oil company, Bridas Corps., which has oil and gas exploration operations in Argentina, Bolivia and Chile.
January 2010	CNPC and others	Iraq	Won the contract to operate the Halfaya oil field in consortium with Total, Petronas, and Iraq's Missan oil.
November 2009	CNPC & BP	Iraq	Inked a technical service contract to expand production of Iraq's Rumaila oil field.
October 2009	CNOOC	Gulf of Mexico	Purchased partial share of Norwegian's Statoil's US assets in deepwater areas of Gulf of Mexico for 100 million.
October 2009	CIC	Russia	Purchased 45% share in Nobel Oil Group for $300 million to fund Russian expansion plans.

Date	Company	Country/ Region	Description
September 2009	CIC	Unknown	Purchased 11% stake in KMG for $939 million by purchasing global depository receipts.
September 2009	Xinjiang Guanghui Investment	Kazakhstan	Purchased Kazakhstan TBM's 49% share for $300 million to jointly develop Zaysan block in eastern Kazakhstan.
September 2009	CNPC/ PetroChina	Alberta, Canada	Purchased 60% of Athabasca Oil Sands Corps' Mackay River and Dover oil sands projects in Alberta, Canada for $1.9 billion.
August 2009	Sinopec	Nigeria	Acquired an additional 51% interest in Block 3 of the joint development zone (JDZ) between Sao Tome and Nigeria.
August 2009	Sinochem	Syria and Colombia	Purchased 100% Emerald for assets in Syria and Colombia for $878 million.
June 2009	CNPC/ PetroChina	Singapore	Purchased 96% of SPC (Singapore) for $2 Billion.
June 2009	Sinopec	N/A	Purchased 100% of Addax for $8.8 billion.
April 2009	CNPC and KMG	Kazakhstan	Purchased MMG in Kazakhstan assets with KMG for $1.7 Billion.
February 2009	CNPC	Indonesia	Acquired 80% interests in the Madura Block.
January 2009	CNPC	Iran	Signed an agreement with National Iranian Oil Company to jointly develop Iran's North Azadegan Oil field for $1.76 billon, which has estimated reserves of 6 billion barrels.
2009	CNOOC and Sinopec	Angola	Purchased 20% stake of Block 32 (Angola) from Marathon Oil for $1.3 billion.
2009	Sinopec	Syria	Purchased 100% of Tanganyika for assets in Syria for $1.8 billion.
2009	CNOOC	Trinidad and Tobago	Acquired a 12.5% interest in the 2C Block and a 12.75% interest in the 3A Block in Trinidad and Tobago.
October 2008	CNPC and Uzbekneftegaz	Uzbekistan	Signed a cooperation agreement to create a joint venture to develop the Mingbulak Oil field.

Date	Company	Country/ Region	Description
June 2008	Sinopec	Nigeria	Acquired 40% interest in oil prospecting license OPL 227.
June 2008	CNPC	Niger	Signed integrated upstream and downstream deals in Agadem block, involving oil field exploration and development, construction and operation of a long-distance pipeline and a joint venture refinery.
June 2008	Sinopec	Australia	Purchased 60% of Australia's AED oil assets for $561 million.
May 2008	PetroChina and PDVSA	Venezuela	Agreed to build a refinery in Guangzhou, China.
2008	CNOOC	Indonesia	Purchased 50% interest in Husky (Madura) Energy's assets for $125 million.
2008	Sinochem	Yemen	Purchased 100% of Soco Yemen for Assets in Yemen for $450 million.
2008	CNOOC	Equatorial Guinea	Offshore exploration.
January 2007	CNPC	Myanmar	Entered into a PSC with Myanmar Oil and Gas Enterprise (MOGE) and acquired oil and gas exploration and production licenses for three deep-water blocks—AD-1, AD-6 and AD-8.
January 2007	CNPC	Venezuela	Signed an agreement with Venezuela's Ministry of Energy and Mines to jointly develop and produce 20 million metric tons of heavy oil annually. On November 6, 2007, CNPC and PDVSA signed a MOU to extend integrated upstream and downstream cooperation in the Orinoco heavy oil belt.
2007	CNPC	Iraq	Reconfirmation of 1997 deal to develop al-Ahdab Oil field in exchange for Chinese cancellation of large portion of Iraqi debt to China.
December 2006	CNOOC	Nigeria	Agreed to pay $2.3 billion for a 45% working interest in Nigerian Oil Mining License (OML) 130 from South Atlantic Petroleum.
October 2006	CNPC and Rosneft	Russia	Jointly established Vostok Energy Ltd., in which CNPC has a 49% stake.

Date	Company	Country/ Region	Description
August 2006	CNPC and others	Uzbekistan	Signed an agreement on the establishment of an investors' consortium comprising of CNPC, Uzbekistan's Uzbekneftegaz, Russia's Lukoil, Malaysia's Petronas and South Korea's KNOC signed a product sharing contract on the exploration and development of oil and gas deposits in the Aral Sea.
August 2006	Sinopec and OVL	Colombia	Jointly acquired with India's OVL 50% of Omimex de Colombia, a subsidiary of U.S. Omimex Resources.
May 2006	CNPC	Equatorial Guinea	Signed a Simplified Purchase Agreement (SPA) for Block M with Fruitex.
April 2006	CNOOC	Kenya	Signed oil exploration contract to prospect for oil.
2006	CNPC/ PetroChina	Chad	Purchased 100% of Block H in Chad from Swiss company Cliveden.
2006	Sinopec	Russia	Purchased 97% of Udmurtneft for assets in Russia for $1.7 billion.
2006	CITIC Resources Holdings	Kazakhstan	Purchased 50% of JSC Karazhanbasmunai for assets in Kazakhstan for $500 million.
2006	CITIC Resources Holdings	Indonesia	Purchased 51% of Seram Block through acquiring assets from KUFOEC for $97.5 million.
2006	Governments	Saudi Arabia	Agreement to set up 10 million cubic meter oil storage facility on Hainan Island.
2006	CNPC	Nigeria	Reached an oil and gas cooperation agreement with the government of Nigeria and won the tender of four blocks, namely OPL298, OPL471, OPL721 and OPL732.
December 2005	CNPC/ KazMunai Gas	Kazakhstan	Completed Atasu-Alashankou oil pipeline; completely operational in July 2006.
December 2005	CNPC	Libya	Signed a risk exploration contract for Block 17-4 with Libya's National Oil Corporation (NOC).

Date	Company	Country/ Region	Description
December 2005	CNPC	Peru	Signed a concession agreement to explore, develop and produce crude oil in two zones (Blocks 111 and 113) in the southeastern region of Madre de Dios.
December 2005	CNPC (50%) and ONGC	Syria	Purchased 38% of Al Furat Production Company from PetroCanada for $574 million.
October 2005	CNPC	Peru	Acquired Block 7.
October 2005	CNOOC and CNPC	Kazakhstan	Signed an MOU with KazMunaiGaz to explore the offshore Darkhan oil field.
October 2005	CNPC	Kazakhstan	Acquired PetroKazakhstan.
September 2005	CNPC	Sudan	Offshore exploration & production of block 15.
September 2005	CNPC and Sinopec	Ecuador	Purchased 100% of EnCana for oil and pipeline interest for $1.47 billion.
August 2005	Sinopec	Angola	Secured a 30% share in Block 3/5.
July 2005	CNPC	Uzbekistan	Agreed to a $600 million oil venture to invest in 23 oil fields.
July 2005	CNPC	Turkmenistan	Signed an agreement on oil and gas cooperation; extended a loan of $24 million for development of oil & gas industry.
June 2005	CNPC	Oman	Signed a Memorandum of Understanding on the investment and cooperation in oil and gas development and downstream operation with Oman Oil Company.
June 2005	CNPC	Mauritania	Owns 65% stake in onshore Block 20 for exploration & production; 100% share of Blocks 12, 13, & 21 for exploration.
May 2005	CNPC	Iran	Won the tender for Block 3.
May 2005	SinoCanadian Petroleum (a subsidiary of Sinopec)	Canada	Agreed to purchase 40% interest in Synenco Energy Inc.'s Northern Lights oil sands project for $83 million.
April 2005	CNOOC	Canada	Paid $122 million for a 16.7% interest in Canadian oil sands company MEG Energy Group.

Date	Company	Country/ Region	Description
April 2005	CNPC	Mongolia	Purchased a 94.4% stake in three petroleum blocks (19, 21 and 22) from SOCO International PLC (UK).
March 2005	Sinopec	Congo-Brazzaville	Two oil exploration & production blocks.
January 2005	CNPC	Indonesia	Received the 25% interests of Bangko Block from El Paso.
January 2005	CNPC	Kazakhstan	Acquired a 100% in ADM, which has exploration licenses in Aryss & Blinov blocks.
January 2005	Sinopec	Cuba	Signed a joint exploration agreement and production agreement with Cubanpetroleo Enterprise.
2005	Sinopec	Canada	Purchased 50% interest in Northern Lights oil sands project for $50 million.
2005	CNOOC	Alberta, Canada	Purchased 14.52% stake in MEG Energy for oil sand business for $220 million.
2005	PetroChina	Nigeria	Secured a one-year supply contract, 30,000 b/d. Four oil exploration blocks reward for stake in Kaduna refinery.
2005	CNPC	Venezuela	Signed an agreement with Petroleos de Venezuela (PDVSA) to jointly explore the Zumano Oil fields in Eastern Venezuela.
December 2004	Sinopec	Nigeria	Signed an agreement with Nigeria Petroleum Development Corp. to develop oil production in 2 blocks in Niger delta (OML 64 & 66).
December 2004	Sinopec	Sudan	Adar-Port Sudan pipeline.
December 2004	Sinopec	Cote d'Ivoire	Became a partner in the San Pedro block.
November 2004	CNPC and NORINCO	Kazakhstan	Jointly acquired 50% of Konys & Bektas oil field.
November 2004	Sinopec	Nigeria	Obtained 2 blocks in Lake Chad Basin.
October 2004	CNPC	Tunisia	Officially acquired the NK exploration block and a 50% holding in the SLK Oil field from Kufpec, a Kuwaiti company.

Date	Company	Country/ Region	Description
October 2004	Sinopec	Iran	Signed an MOU for a 25-year $70 billion agreement to import LNG in exchange for developing Yadavaran Oil field.
October 2004	Sinopec	Angola	Secured a 50% interest in Angola's Block 18.
October 2004	Sinopec	Mali	Obtained 2 exploration blocks.
October 2004	Sinopec	Kazakhstan	Purchased $100 million worth of Kazakh oil assets from First International Oil Company, a small U.S. firm.
August 2004	CNPC	Niger	Obtained 2 exploration blocks: Bilma and Tenere.
July 2004	CNPC and Sinopec	Algeria	Granted with 3 exploration blocks. Acquired 100% of Block 438b.
June 2004	CNPC	Sudan	Adar/Yale fields, 300,000 b/d.
June 2004	Sinopec	Yemen	Awarded rights to explore blocks 69 in Shabwah and 71 in Hadramout.
May 2004	Sinopec	Brazil	Signed a strategic agreement of cooperation.
March 2004	CNPC	Yemen	Energy agreements signed for mutual exploration of Yemeni oil fields by CNPC and Yemeni National Oil Company.
February 2004	CNPC	Kazakhstan	Conveyed a 50% stake of North Buzachi, Inc. to Canadian company Nelson Resources for $90 million.
December 2004	CNPC	Venezuela	Signed an agreement with Petroleos de Venezuela (PDVSA) to jointly explore the Zumano Oil fields in Eastern Venezuela.
January 2004	Sinopec	Azerbaijan	Signed a 25-year $220 million deal to rehabilitate Azerbaijan's Garachukhur oil field.
January 2004	CNPC	Peru	Purchased Block 8 and Block 1AB from Argentine Pluspetrol for $200 million.
2004	Sinopec	Angola	Signed a memorandum of understanding for joint refinery & off-shore prospecting.
2004	CNPC and others	Iraq	Signed a 22-year production-sharing contract to develop al-Ahdab field for $1.3 billion.

Date	Company	Country/ Region	Description
2004	Sinopec	Gabon	Exploration of 3 offshore blocks; 2 onshore blocks.
2004	PetroChina	Indonesia	Has a 25% interest in, and operates, Sukawati oil field.
2004	Sinopec and CNOOC	Myanmar	Awarded with four oil exploration blocks.
2004	Sinopec	Saudi Arabia	Agreement to upgrade and triple capacity of Quanzhou refinery in Fujian Province in cooperation with Saudi Aramco and ExxonMobil valued at $3.5 billion; ground broken July 2005.
2004	Government of China (Chinese EXIM Bank)	Angola	Provided $2 billion line of credit for 10,000 b/d supply.
December 2003	CNPC	Algeria	Invested $31 million over 3 years to prospect for oil. Has 75% equity Blocks 112/102a & 350.
October 2003	CNPC	Ecuador	Acquired block 11.
October 2003	CNPC	Kazakhstan	Bought out ChevronTexaco's interests of North Buzachi, Inc.
May 2003	CNPC	Kazakhstan	Purchased a 25% interest in AktobeMunaigas Corp. increasing its total interest to 85%.
May 2003	CNPC	Algeria	Won the tender for the Adrar upstream and downstream integrated project with a contract term of 23 years.
April 2003	PetroChina	Indonesia	Purchased a 45% interest in the operation of Block Jabung.
March 2003	CNPC	Syria	Signed a development and production contract on Gbeibe Oil field with the Syrian Ministry of Petroleum and Mineral Resources and the Syrian Petroleum Corporation (SPC).
January 2003	CNPC	Azerbaijan	Acquired 50.26% equity stake of Gobustan oil field.
2003	Sinochem	Ecuador	Purchased 14% interest in Block 16 in Ecuador from ConocoPhilips for $100 million.
October 2002	CNPC	Indonesia	Acquired 45% equity of Mineral Resources International Inc. (Malaysia) in Block SP.

Date	Company	Country/ Region	Description
April 2002	CNPC	Indonesia	Purchased Devon Energy Corporation for six blocks in Indonesia for $585 million.
April 2002	CNPC	Oman	Acquired 50% of Block 5—Daleel and Mezoon with Oman Oil Company SAOC (OOC).
January 2002	CNPC	Turkmenistan	Signed an EOR contract for the field with Turkmenneft and acquired a 100% stake for five years.
January 2002	CNPC	Azerbaijan	Signed a production contract and acquired 50% equity rights of the Kursangi & Karabagli oil field.
2002	CNOOC	Indonesia	Purchased YPF Repsol's assets (Southeast Sumatra etc) for $585 million.
2002	CNPC/ PetroChina	Indonesia	Purchased Devon Energy Corporation for six blocks in Indonesia for $585 million.
December 2001	CNPC	Myanmar	Entered into a 20 year contract of improving oil recovery with Myanmar Oil and Gas Enterprise (MOGE) on Block IOR-4, which situated in southern Myanmar.
November 2001	CNPC	Myanmar	CNPC purchased the Bagan project in the center Myanmar from TG World and acquired a 100% interest of Block IOR-3, TSF-2 and RSF-3.
November 2000	CNPC	Sudan	Won the tender for Block 3/7 owning a 41% stake.
June 2000	CNPC	Sudan	Khartoum refinery, 70,000 b/d.
1999	CNPC	Sudan	Heglig-Port Sudan Pipeline (500,000 bpd)—A pipeline from fields to the Red Sea.
June 1997	CNPC	Venezuela	Won tenders to develop the Caracoles and Intercampo Oil fields.
June 1997	CNPC	Kazakhstan	Purchased 60% of AktobeMunaiGas Production Association for $43 billion.
March 1997	CNPC	Sudan	Won the tender for Block 1/2/4 and jointly established GNPOC together with Petronas, SPC and Sudapet in June the same year.

Date	Company	Country/ Region	Description
1997	CNPC	Sudan	Acquired a 40% stake in the Greater Nile Petroleum Operation Company consortium to explore & develop the Heglig & Unity fields.
September 1995	CNPC	Sudan	Acquired the development rights of Block 6.
January 1994	CNPC	Peru	Acquired Block 6.
1993/1995	CNOOC	Indonesia	Purchased 32.5% interest in an Oil field in the Straits of Malacca, and an additional 6.93% in 1995.
April 1993	CNPC	Canada	Acquired a 15.8865% operating interest in North Twining Oilfield.
1993	CNPC	Thailand	Acquired rights and interests in the Banya development block, BYW-NS development block and L21/43 risk exploration block.
Recent Years	China	Turkmenistan	Extended three $12-million low interest loans to Turkmenistan for the purchase of Chinese oil rigs.
Undetermined	Government of China	Kenya	Potential for a joint venture with Kenya Pipeline Company to building a $1.4 billion pipeline connecting oil fields in Unity province in Sudan to the port city of Lamu, Kenya.

Sources: Julie Jang and Jonathan Sinton, Overseas *Investments By Chinese National Companies: Assessing the Drivers and Impacts*, International Energy Agency, February 2011, 39–40; CNPC; U.S. Department of Energy; GlobalData, Eurasia Group, *CNOOC Annual Report 2010* and John W. Garver.

Appendix C

INDIAN FOREIGN OIL ACQUISITIONS AND EXPLORATION AND PRODUCTION PROJECTS

Date	Company	Country	Description
April 2011	OVL and KMG	Kazakhstan	OVG signed definitive agreements for acquisition of 25% participating interest in Satpayev exploration block with KazMunaiGas, the national oil company of Kazakhstan.
December 2010	OVL, Petrobras, and Ecopetrol	Brazil	Acquired 43.5% stake in Block S-74.
May 2010	Indian Oil Company (IOC), OVL, OIL, CVP, Repsol YPF, Petronas	Venezuela	Purchased a 40% share (OIL-3.5%, OVL-11%, Petronas-11%, IOC-3.5% and Repsol YPF-11%) in an "Empresa Mixta" (or "Mixed Company") which will develop the Carabobo 1 Norte and Carabobo 1 Centro blocks located in the Orinoco Heavy Oil Belt.
March 2009	IOC, Medco Energi, Kuwait Energy and OIL.	Yemen	Consortium acquired a 100% interest (IOC-15%, Medco-45%, Kuwait Energy-25% and OIL-15%) to develop Block 82.
March 2009	IOC, Medco Energi, Kuwait Energy and OIL.	Yemen	Consortium acquired a 100% interest (IOC-15%, Medco-45%, Kuwait Energy-25% and OIL-15%) to develop Block 83.
December 2008	OVL and Pacific Stratus Energy Colombia	Colombia	Acquired a 50% stake in Block SSJN-7.

Date	Company	Country	Description
December 2008	OVL and Petrodorado	Colombia	Acquired a 70% stake in Block CPO-5.
August 2008	OVL and Petrobras	Brazil	Acquired a 25% stake in Blocks BM-SEAL-4 & BM-BAR-1.
August 2008	OVL	Russia	Set-up a wholly-owned subsidiary Jarpeno Ltd on 12th August, 2008 in Cyprus. Jarpeno Limited has acquired Imperial Energy in January, 2009 which through its subsidiaries owns the licenses of the blocks/ fields in Russia.
June 2008	IOC, OIL, and Reliance Exploration & Production	Timor Leste	Consortium acquired a 100% interest (IOC-12.5%, OIL-12.5%, Reliance-75%) to develop Block K.
June 2008	IOC, OIL and Sonatarch	Libya	Consortium acquired a 100% exploration and production interest (Sonatarch-50%, IOC-25% and OIL-25%) in Area 95/96 (Blocks 21, 22 & 24).
April 2008	OVL and CVP	Venezuela	OVL acquired 40% stake in San Cristobal Project.
March 2008	OVL	Brazil	Acquired 100% stake in Blocks BM-ES-42 and BM-S-73. Farmed out 43.5% to Petrobras and 13% to Ecopetrol in 2009-2010.
2008	OVL, Exxon-Mobil, Sodeco, SMNG, and RN Astra.	Russia	Holds a 20% equity stake the Exxon-Mobil-led consortium that operates the Sakhalin-I project in Russia. The oil fields associated with Sakhalin-I hold recoverable crude oil reserves of 2.3 billion barrels.
2008	GSPC, OIL and IOC	Egypt	Consortium (GSPC-50%, OIL-25% and IOC-25%) led by GSPC awarded the contract to develop Block 3 and 4.
September 2007	OVL and Ecopetrol	Colombia	Acquired a 50% stake in Blocks RC-9 and RC-10.
September 2007	OVL, Petrobras and Ecopetrol	Colombia	Acquired a 40% interest in Blocks RC-8.
September 2007	OVL	Myanmar	Acquired three offshore deepwater exploration Blocks: AD-2, AD-3 and AD-9.
March 2007	OVL	Libya	Signed a contract for a 100% stake to develop Block 43.

Date	Company	Country	Description
February 2007	ONGC Mittal Energy Limited (OMEL), EMO and Total	Nigeria	Consortium acquired 100% interest (OMEL-45.5%, EMO-40% and Total-5%) to develop Block OPL 279.
February 2007	ONGC Mittal Energy Limited (OMEL), EMO and Total	Nigeria	Consortium acquired 100% interest (OMEL-64.33%, EMO-10% and Total-25.67%) to develop Block OPL 285.
September 2006	OVL	Cuba	Signed a Production Sharing Contract for Blocks N-34 and N-35 with 100% PI in Cuba's Exclusive Economic Zone.
September 2006	OVL (OAAL) and Sinopec	Colombia	Jointly acquired with India's ONGC 50% of Omimex de Colombia, a subsidiary of U.S. Omimex Resources.
August 2006	IOC, OIL, Sunteras Resources Ltd., and Summit Oil Intl. Ltd.	Nigeria	Consortium acquired a 100% interest (OIC-17%, OIL-17%, Suntera-35% and Summit-30%) to develop Block OPL 205.
May 2006	OVL, Repsol YPF and Stat Oil	Cuba	Acquired a 30% stake in Blocks N-25, 26, 27, 28, 29 & N-36.
May 2006	OVL	Vietnam	Acquired 100% interest in developing Blocks 127 and 128.
April 2006	IOC, OIL and Marvis Pte Ltd.	Gabon	Consortium acquired 100% interest (IOC-45%, OIL-45% and Marvis-10%) to develop Block Shakthi.
April 2006	OVL, Shell, and Petrobras	Brazil	Acquired a 15% stake in Block BC-10.
March 2006	OVL, Daewoo, KOGAS, GAIL and MOGE	Myanmar	Acquired a 17% stake in Block A-3.
December 2005	OVL and CNPC	Syria	Purchased 38% of Al Furat Production Company from PetroCanada for $574 million.
December 2005	OVL	Libya	Acquired 100% interest to develop Block 81-1.
December 2005	IOC and OIL	Libya	Consortium acquired a 100% exploration and production interest (IOC-50% and OIL-50%) in Block 102/104.
August 2005	OIL and OVL	Sudan	Completed an oil pipeline from Al Rawyan to Port Sudan. Shares: OIL-10% and OVL-90%.

Date	Company	Country	Description
March 2005	IOC and OIL	Libya	Consortium acquired a 100% exploration and production interest (IOC-50% and OIL-50%) in Area 86.
May 2004	OVL, Petronas, and Sudapet	Sudan	Consortium acquired 100% interest (OVL-24.125%, Petronas-67.875% and Sudapet-8%) to develop Block 5A.
January 2004	OVL, IPR International, and Tri Ocean Mediterranean	Syria	Acquired a 60% share in Block 24.
June 2003	OVL and Turkish Petroleum Overseas Company	Libya	Consortium acquired 100% interest (OVL-49% and Turkish Petroleum-51%) to develop Block NC-189. All wells were plugged and abandoned in March 2010.
March 2003	OVL, CNPC, Petronas and Sudapet	Sudan	Purchased 25% equity stake in the Greater Nile Petroleum Operating Company (GNPOC). The GNPOC acreage in Sudan holds proved crude oil reserves of more than 1 billion barrels with current production levels at roughly 300,000 bbl/d from 10 fields. GNPOC also operates a 935-miles crude oil pipeline that pumps oil to Port Sudan Export. Shares: CNPC-40%, Petronas-30% and Sudapet-5%.
January 2003	OVL, BP and PetroVietnam	Vietnam	Part of the BP-led consortium that developed Block 06.1 offshore Vietnam. OVL's development share is 244 million as of 31 March 2011. Shares: OVL-45%, BP-35% and PetroVietnam-20%.
December 2002	IOC, OVL and OIL	Iran	Consortium acquired a 100% exploration interest (IOC-40%, OVL-40% and OIL-20%) in Block Farsi.
April 2002	OVL, Daewoo, KOGAS, GAIL and MOGE	Myanmar	Acquired a 17% stake in Block A-1.
May 2001	OVL	Iraq	Acquired a 100% stake in Block 8.
Unknown	Reliance Industries	International	Has stakes in 13 international blocks: Peru (2), Yemen (3), Oman (2), Kurdistan (2), Colombia (2), East Timor (1) and Australia (1).

Sources: Energy Information Administration, U.S. Department of Energy, County Analysis Briefs: India, August 2010; CNPC, ONGC (OVL), IOC, OIL, Reliance.

Appendix D

JAPANESE FOREIGN OIL ACQUISITIONS AND EXPLORATION AND PRODUCTION PROJECTS

Date	Company	Country/ Region	Description
May 2011	Mitsui E&P Middle East B.V.	Oman	Acquired a 35% interest in Blocks 9 and 27.
April 2011	Idemitsu Oil & Gas Co., Ltd. (Idemitsu Norge)	Norway	Acquired licenses to explore in Blocks 6601/6, 9, 6602/4, 7 (PL598), 6605/2, 3, 6, 6606/1, 6706/10 (PL599), 7220/6, 9, 11, 12, 7221/4 (PL609) and 7324/9, 7325/7 (PL614).
January 2011	Idemitsu Oil & Gas Co., Ltd. (Idemitsu Norge)	Norway	Acquired licenses to explore in Blocks 35/7 (PL377BS), 35/6(PL578), 6202/8, 9, and 6203/7(PL580).
December 2009	Japex Garraf Ltd.	Iraq	Obtained a license to explore in the Garraf oilfield.
November 2009	Japan CBM Ltd. (includes Japex and JX Nippon Oil)	Indonesia	Signed a production contract for Sanga Sanga CBM Block (onshore East Kalimantan).
November 2009	Inpex Alpha, Ltd.	Western Australia	Granted a production license in Block WA-43-L.
October 2009	JX Nippon Oil Exploration U.S.A.) Ltd.	United States (Gulf of Mexico)	Discovered oil and gas in Block WC/265/266.
September 2009	INPEX Ichthys Pty Ltd.	Western Australia	Obtained a retention lease for Block WA-37-R.
August 2009	JX Nippon Oil & Gas Exploration Corporation	East Timor	Acquired 15% working interest in JPDA 06-103 Block.

Date	Company	Country/ Region	Description
March 2009	INPEX Alpha, Ltd.	Western Australia	Acquired a working interest in Block WA-255-P.
February 2009	Cosmo Oil Ashmore Ltd.	Australia	Has mining claims in Block AC/RL5.
February 2009	Cosmo Energy Exploration & Development Co., Ltd. & Cosmo Oil Ashmore Ltd.	Australia	Has mining claims in Blocks AC/P17 and AC/RL6.
February 2009	Cosmo Energy Exploration & Development Co., Ltd.	Australia	Has mining claims in Blocks AC/P4 and AC/P32.
2009	Universe Gas & Oil Company, Inc.	Indonesia	Acquired an interest in Sanga Sanga Block (onshore East Kalimantan).
2009	Energi Mega Pratama Inc.	Indonesia	Acquired a 100% interest in Kangean Block (offshore East Java).
October 2008	Inpex Alpha, Ltd.	Western Australia	Granted a production license in Block WA-35-L.
June 2008	JX Nippon Oil & Gas Exploration Corporation	Western Australia	Acquired working interest in Block WA-412-P.
May 2008	Japan Canada Oil Sands Ltd. (Japex subsidiary)	Canada	Acquired a 100% interest in the Hangingstone Block.
March 2008	Idemitsu Cuu Long Petroleum Co., Ltd.	Vietnam	Acquired a 15% share of Block 09-3.
2007/2008	Cosmo Energy Exploration & Development Co., Ltd.	Qatar	Has mining claims to Blocks 3 and 11.
December 2007	JX Nippon Oil & Gas Exploration Corporation	Thailand	Acquired an interest in Block B6/27.
November 2007	INPEX Canada, Ltd.	Canada	Acquired a 10% interest in the Joslyn Oil Sands Upstream Project.
November 2007	JX Nippon Oil & Gas Exploration Corporation	Vietnam	Acquired an interest in Block 16-2.
October 2007	Mitsui Oil Exploration Co. Ltd.	Thailand	Extended the production period to 2022 for four offshore blocks (10-13).

Date	Company	Country/ Region	Description
July 2007	Mitsui E&P Australia Pty Ltd.	New Zealand	Acquired a 35% interest in Tui Area Oilfield offshore North Island of New Zealand.
February 2007	JX Nippon Exploration and Production (U.K.) Ltd.	United Kingdom (North Sea)	Acquired interests in Blocks 15/23c, 15/24a, 15/28a, 15/29e, 22/25c and 30/3b.
January 2007	Japex Buton Ltd.	Indonesia	Signed a PSC for the Buton Block, an onshore/offshore block on Buton Island, Southeast Sulawesi.
2007	Sakhalin Oil and Gas Development Co., Ltd. (Japex subsidiary)	Russia	Acquired an interest in the Sakhalin I Project (Chayvo, Odoptu and Arkutun-Dagi (offshore Sakhalin).
Since 2007	Idemitsu Oil & Gas Co., Ltd. (Idemitsu Petroleum UK Ltd.)	United Kingdom	Holds interests in a total of 31 licenses in the UKCS which include nine producing fields in the Central North Sea, as well as several discoveries, and exploration blocks both in the North Sea and West of Shetland.
2007	Cosmo Oil Co. Ltd.	Qatar	Approved the development plan for A-Structure South Field under the auspices of Qatar Petroleum Development Co. Ltd.
July 2006	INPEX Alpha, Ltd.	Western Australia	Acquired a working interest in Block WA-357-P.
April 2006	MitEnergy Upstream LLC	United States (Gulf of Mexico)	Acquired a 50% interest in the assets of Pogo Producing Company located offshore.
April 2006	Teikoku Oil (North America) Co., Ltd.	United States (Gulf of Mexico)	Participated in oil and gas development projects in the shallow waters of the U.S. Gulf of Mexico.
December 2005	JX Nippon Oil & Gas Exploration Corporation	Western Australia	Acquired working interest in Block WA-290-P.
December 2005	Japex Libya Ltd., JX Nippon Oil & Gas Exploration Ltd., and Mitsubishi Corporation	Libya	Acquired an interest in Block 40-3/4.

Date	Company	Country/ Region	Description
December 2005	JX Nippon Oil & Gas Exploration and Mitsubishi Corporation	Libya	Acquired an interest in Block 2-1/2.
September 2005	JX Nippon Exploration and Production (U.K.) Ltd.	United Kingdom (North Sea)	Acquired interests in Blocks 16/3d, 210/18, 210/19a and 210/20a.
July 2005	JJI S&N B.V.	Iran	Obtained a 20% interest to develop the Soroosh and Nowrooz oilfields.
June 2005	Mitsui Oil Exploration Co. Ltd.	Thailand	Acquired a 46.3% interest Blocks B8/32 and 9A.
2005	Japex Libya Ltd.	Libya	Acquired a 100% interest in 176/4 Block.
October 2004	Idemitsu Oil & Gas Co., Ltd., JX Nippon Oil & Gas Exploration Corporation and Teikoku Oil (Con Son) Co., Ltd.	Vietnam	Entered a production sharing contract with Vietnam Oil and Gas Group (Petrovietnam) for the Blocks 05-1b and 05-1c offshore Vietnam.
March 2004	Mitsui E&P Australia Pty Ltd.	Western Australia	Acquired a 40% interest in Blocks WA-28-L and WA-271-L.
2004	Teikoku Oil de Burgos, S.A.de C.V.	Mexico (Gulf of Mexico)	Participated in gas development and production operations in the Cuervito and Fronterizo Blocks located in the Burgos Basin, Mexico.
April 2003	INPEX Southwest Caspian Sea, Ltd.	Azerbaijan	Acquired a 10% working interest in the Azeri-Chirag-Gunashli (ACG) oil fields.
August 2002	JX Nippon Exploration and Production (U.K.) Ltd.	United Kingdom (North Sea)	Acquired an interest in the Brae Oilfields.
2002	Idemitsu Oil & Gas Co., Ltd. (Idemitsu Norge)	Norway	Acquired a 15% share of the Fram fields (Block 35/11, PL090).
November 2001	Teikoku Oil Algeria Co., Ltd.	Algeria	Acquired a 10.29% working interest in the El Ouar I and II onshore blocks.
October 2001	Inpex Corporation, Mitsubishi, JX Nippon Oil Exploration Corporation	Indonesia	Acquired a 22.9% interest of the Berau Block as joint venture (MI Berau B.V.) with Mitsubishi. Produces crude oil.

Date	Company	Country/ Region	Description
July 1999	INPEX Alpha, Ltd.	Western Australia	Acquired a working interest in Block WA-155-P (Part I) (Van Gogh and Ravensworth).
July 1999	Frade Japão Petróleo Limitada (FJPL)	Brazil	Acquired a 12.75% working interest in the Frade block in Brazil's offshore Northern Campos basin.
December 1998	INPEX Masela, Ltd.	Indonesia	Acquired 100% interest in the Masela (Abadi) Block.
September 1998	INPEX Browse, Ltd.	Western Australia	Acquired a working interest in Blocks WA-285-P, WA-274-P, WA-281-P, WA-341-P, WA-343-P, WA-344-P, WA-410-P and WA-411-P.
August 1998	INPEX North Caspian Sea, Ltd.	Kazakhstan	Acquired a working interest in the Offshore North Caspian Sea block (Kashagan Oilfield) inside territorial waters.
October 1997	Idemitsu Oil & Gas Co., Ltd.	Cambodia	Acquired a 33% share of Areas I & II offshore Cambodia.
September 1997	Cosmo Oil Co. Ltd.	Qatar	Qatar Petroleum Development Co., Ltd. (subsidiary of and 75% owned by Cosmo) acquired concessions to explore the Al-Karkara oil field and the A-Structure North oil field.
May 1997	JX Nippon Oil & Gas Exploration Corporation	Western Australia	Acquired working interest in Block WA-191-P.
1997	JX Nippon Oil & Gas Exploration Corporation	Papua New Guinea	As a shareholder of Murray Petroleum (includes INPEX, Japan Oil, Gas and Metals National Corporation and Japan Petroleum Exploration Corporation Ltd.), acquired a petroleum prospecting license 190 (PPL 190).
December 1996	JX Nippon Exploration and Production (U.K.) Ltd.	United Kingdom	Acquired interest in Magnus Oilfield.
March 1996	JX Nippon Oil & Gas Exploration Corporation	Papua New Guinea	As a shareholder of Southern Highlands Petroleum (includes INPEX and Japan Petroleum Exploration Corporation Ltd.), acquired a petroleum development licenses 56 and 161 (PPLs 56 and 161) for the Gobe Oilfield.

Date	Company	Country/ Region	Description
July 1994	INPEX Alpha, Ltd.	Western Australia	Acquired working interests in Blocks WA-155-P (Part II) (Van Gogh and Ravensworth) and WA-12-L (Deep) offshore.
October 1992	JX Nippon Oil & Gas Exploration Corporation	Vietnam	Acquired an interest in Block 15-2.
July 1992	Teikoku Oil and Gas Venezuela, C.A.	Venezuela	Acquired a 100% working interest in the Guarico Oriental Block.
February 1992	JX Nippon Oil & Gas Exploration Corporation	Canada	Acquired an interest in the Syncrude Project.
January 1992	INPEX Timor Sea, Ltd.	East Timor Sea	Acquired an interest in Block JPDA06-105 (Kitan Oilfield).
July 1990	JX Nippon Oil & Gas Exploration Corporation and Mitsubishi Corporation	Papua New Guinea	Acquired working interests in Blocks PPL219/239/260 and offshore Blocks PPL234/244 through purchase of shares (100%) of Merlin Petroleum Company.
1990	Idemitsu Oil & Gas Co., Ltd. (Idemitsu Norge)	Norway	Acquired a 9.6% interest in the 2 blocks (34/4 & 34/7) including the Snorre field.
Since 1990	Idemitsu Oil & Gas Co., Ltd. (Idemitsu Norge)	Norway	Acquired interests in Tordis/Vigdis Fields, Statfjord East Field, Sygna Field, and Knarr Fields (Block 34/3, PL373S).
February 1989	INPEX Alpha, Ltd.	Western Australia	Acquired a 20% interest in WA-210-P.
July 1977	Inpex Corporation	Indonesia	Acquired in the South Natuna Sea Block B.
June 1975	Egyptian Petroleum Development Co., Ltd. (Inpex and Mitsui & Co.)	Egypt	Acquired a 100% working interest in the West Bakr Block.
February 1973	Japan Oil Development Co., Ltd. (JODCO) (Inpex subsidiary)	United Arab Emirates	Acquired an interest in the ADMA block located offshore Abu Dhabi.
1972	Idemitsu Oil & Gas Co., Ltd., and Mitsui Oil Exploration Co., Ltd. (MOECO)	Thailand	Acquired a 50% share of Blocks 5 and 6 offshore Thailand.

Date	Company	Country/ Region	Description
July 1970	Teikoku Oil (D.R. Congo) Co., Ltd.	Democratic Republic of the Congo	Acquired a 17.03% working interest in July 1970 in an offshore area of the DRC.
1970	Cosmo Oil Co. Ltd., JX Nippon Oil & Gas Exploration, and Mitsui Oil Exploration Co. Ltd.	United Arab Emirates/ Qatar	Won the concession to explore in Al Bunduq oil field as the equity-affiliate, United Petroleum Development Co., Ltd.
1969	Cosmo Oil Co. Ltd. and Japan Energy Corporation (JX Nippon)	United Arab Emirates	Formed a consortium (Abu Dhabi Oil Co., Ltd.) to explore for oil in the Mubarraz, AR and GA oil fields.
October 1966	Inpex Corporation	Indonesia	Acquired 50% share of the Offshore Mahakam Block and Attaka Unit

Sources: Cosmo Oil Co. Ltd.; Idemitsu Oil & Gas Co., Ltd.; Inpex Corporation; Mitsui; and JX Nippon Oil & Gas Exploration Corporation.

Selected Bibliography

Al-Suwaidi, Jamal S. "Introduction." In *China, India and the United States: Competition for Energy Resources*. Abu Dhabi, UAE: Emirates Center for Strategic Studies and Research, 2008.

Alterman, Jon B., "China's Soft Power in the Middle East." In *Chinese Soft Power and Its Implications for the United States*, edited by Carola McGiffert, 63–76. Washington, DC: CSIS Press, March 2009.

Alterman, Jon B., and John W. Garver. *The Vital Triangle: the United States and the Middle East*. Washington, DC: CSIS Press, 2008.

Andrews-Speed, Philip, and Roland Dannreuther. *China, Oil and Global Politics*. London: Routledge, 2011.

Bajpaee, Chietigj. "Quest for Energy: The Role of Consumer Governments and National Oil Companies." in *China, India and the United States: Competition for Oil Resources*, 77–111. Abu Dhabi, UAE: Emirates Center for Strategic Studies and Research, 2008.

Bhalla, Madhu. "Geopolitics of Economic Relations." In *China–Pakistan Strategic Cooperation: Indian Perspectives*, edited by Swaran Singh, 251. New Delhi, India: Manohar Publishers, 2007.

Blanchard, Christopher M., Nicolas Cook, Kerry Dumbaugh, Susan B. Epstein, Shirley A. Kan, Michael F. Martin, Wayne M. Morrison, Dick K. Nanto, Jim Nichol, Jeremy M. Sharp, Mark P. Sullivan, and Bruce Vaughn. *Comparing Global Influence: China's and U.S. Diplomacy, Foreign Aid, Trade, and Investment in the Developing World*, CRS Report RL34620, Washington, DC: Library of Congress, Congressional Research Service, August 15, 2008.

Bliss, Catherine E., "China's Projection of Soft Power in the Americas." In *Chinese Soft Power and Its Implications for the United States*, edited by Carola McGiffert, 45. Washington, DC: CSIS Press, March 2009.

Bush, Richard. *The Perils of Proximity: China–Japan Security Relations*. Washington, DC: Brookings Institution Press, 2010.

Cameron, Fraser, and Yongnian Zheng. "Key Elements of a Strategic Partnership." In *China–EU: A Common Future,* edited by Stanley Crossick and Etienne Reuter, 6. Singapore: World Scientific Publishing, 2007.

Cheru, Fantu, and Cyril Obi. "Introduction—Africa in the Twenty-First Century: Strategic and Development Challenges." In *The Rise of China and India in Africa: Challenges, Opportunities and Critical Interventions,* edited by Fantu Cheru and Cyril Obi, 5. New York: Zed Books, 2010.

Chin, Victor A. "The Prospect of China's Access to Naval Facilities in Burma and the Ramifications for Regional Security." Master's thesis. Naval Postgraduate School, 2007.

Cole, Bernard. *Oil for the Lamps of China–Beijing's 21st Century Search for Energy.* Washington, DC: Institute for National Strategic Studies, National Defense University, 2003.

———. *Sea Lanes and Pipelines: Energy Security in Asia.* Westport, CT: Praeger Security International, 2008.

Cole, J. Michael. "China increases territorial UAVs," *Taipei Times,* September 24, 2012.

Conn, Iain. "Energy Trends and Technologies for the Coming Decades." Paper presented to the Harvard University Center for the Environment, Cambridge, MA, March 8, 2006.

Corrales, Javier. "China and Venezuela's Search for Oil Markets." In *Latin America Facing China: South-South Relations beyond the Washington Consensus,* edited by Alex E. Fernandez Jilberto and Barbara Hogenboom, 124. New York: Berghahn Books, 2012, 124.

Courtney, Hugh, Jane Kirkland, and Patrick Viguerie. "Strategy Under Uncertainty." *Harvard Business Review* (November–December 1997): 68–71.

Davies, Martyn J. "Special Economic Zones: China's Developmental Model Comes to Africa." In *China Into Africa: Trade, Aid and Influence,* edited by Robert I. Rotberg, 137, 142. Washington, DC: Brookings Institution Press, 2008.

DeLisle, Jacques. "Contested Terrain: China's Periphery and International Relations with Asia." *E-Notes,* Foreign Policy Research Institute, March 2012. *www.fpri.org/enotes/2012/201202.delisle.china-periphery.pdf.*

De Paiva, Bruno. "Japan's National Involvement in the Indian Ocean Region." *Future Directions International,* July 29, 2011.

Dormandy, Xenia. "India's Foreign Policy." Paper presented at the India Forum, Madrid, Spain, November 5, 2007.

Downs, Erica Strecker. *China's Quest for Energy Security.* Santa Monica, CA: RAND Corporation, 2000.

Eisenman, Joshua. "China's Post-Cold War Strategy in Africa: Examining Beijing's Methods and Objectives." In *China and the Developing World: Beijing's Strategy for the Twenty-First Century,* edited by Joshua Eisenman, Eric Heginbotham, and Derek Mitchell Armonk, 35–36. New York: Sharpe, 2007.

Ellis, R. Evan. *China–Latin America Military Engagement: Good Will, Good Business and Strategic Position.* Carlisle, PA: Strategic Studies Institute, August 2011.

Evans, Peter C., and Erica S. Downs. *Untangling China's Quest for Oil through State-backed Financial Deals.* Washington, DC: The Brookings Institution (Policy Brief 154), May 2006.

Flanagan, Stephen J. "China's Military Diplomacy and Engagement in Latin America." In *Chinese Soft Power and Its Implications for the United States*, edited by Carola McGiffert, 56. Washington, DC: CSIS Press, March 2009.

Gill, Bates, and Melissa Murphy. *China–Europe Relations: Implications and Policy Reponses for the United States*. Washington, DC: CSIS Press, May 2008.

Gill, Bates, Chin-hao Huang, and J. Stephen Morrison. *China's Expanding Role in Africa: Implications for the United States*. Washington, DC: CSIS Press, January 2007.

Gill, Bates, and Matthew Oresman. *China's New Journey to the West: China's Emergence in Central Asia and Implications for U.S. Interests*. Washington, DC: CSIS Press, August 2003.

Harold, Scott, and Alireza Nader. *China and Iran: Economic, Political and Military Relations*. Santa Monica, CA: RAND Corporation, 2012.

Herberg, Mikkal. "China's Search for Energy Security: the Implications for Southeast Asia." In *China, the United States, and Southeast Asia: Contending Perspectives on Politics, Security, and Economics*, edited by Evelyn Goh and Sheldon W. Simon, 79. New York: Routledge, 2008.

Hille, Kathrin. "Chinese boats fish in dangerous waters." *Financial Times*, April 24, 2012.

Hoffman, Stephen A. "Perception and China Policy in India." In *The India–China Relationship: What the United States Needs to Know*, edited by Francine R. Frankel and Harry Harding, 33–74. Washington, DC: Woodrow Wilson Center Press, 2004.

Horta, Loro. "In Uncle Sam's Backyard: China's Military Influence in Latin America." *Military Review* (September–October 2008): 51.

Huang, Yukon. "The Economics Behind the China–Japan Dispute." *Financial Times*, September 24, 2012. http://carnegieendowment.org/2012/09/24/economics-behind-china-japan-dispute/dwc6.

Inderfurth, Karl, and Bruce Riedel. "Breaking More Naan with Delhi: The Next Stage in U.S.–India Relations." *The National Interest* (Nov/Dec 2007): 56–62.

INPEX Corporation, *Annual Report 2011: A Firm Commitment to Sustainable Energy*, 2011.

Jiang, Julie, and Jonathan Sinton. *Overseas Investments by Chinese Oil Companies: Assessing the Drivers and Impacts*. Organization for Economic Co-operation and Development/International Energy Agency. Paris, France, February 2011.

Jiang, Wenran. "China's Emerging Strategic Partnerships in Africa." In *China Into Africa: Trade, Aid and Influence*, edited by Robert I. Rotberg, 59. Washington, DC: Brookings Institution Press, 2008.

Johnston, Corey S. "Transnational Pipelines and Naval Expansion: Examining China's Oil Insecurities in the Indian Ocean." Master's thesis, Naval Postgraduate School, 2008.

Kaufman, Alison A. *China's Participation in Anti-Piracy Operations off the Horn of Africa: Drivers and Implications*. Alexandria, VA: Center for Naval Analysis, July 2009.

Kaye, Dalia Dassa. *Talking to the Enemy: Track-Two Diplomacy in the Middle East and South Asia*. Santa Monica, CA: RAND Corporation, 2007.

Kemp, Geoffrey. *The East Moves West: India, China, and Asia's Growing Presence in the Middle East*. Washington, DC: Brookings Institution Press, 2010.

Klare, Michael T. "Petroleum Anxiety and the Militarization of Energy Security," In *Energy Security and Global Politics: the Militarization of Resource Management*, edited by Daniel Moran and James A. Russell, 44–46. New York: Routledge, 2009.

———. *Rising Powers, Shrinking Planet: The New Geopolitics of Energy*. New York: Metropolitan Books, 2008.

Large, Daniel, and Luke Patey. "Sudan Looks East." In *Sudan Looks East: China, India and the Politics of Asian Alternatives*, edited by Daniel Large and Luke Patey, 7, 24, 27. Rochester, NY: James Currey, 2011.

Leverett, Flynt. "Resource Merchantism and the Militarization of Resource Management: Rising Asia and the Future of American Primacy." In *Energy Security and Global Politics: the Militarization of Resource Management*, edited by Daniel Moran and James A. Russell, 211, 217. New York: Routledge, 2009.

Liang Guanglie. "Seychelles Chosen for China's First Overseas Naval Base," *Want China Times*, December 13, 2011.

Lieberthal, Kenneth, and Mikkal Herberg. "China's Search for Energy Security: Implications for U.S. Policy." *NBR Analysis* 17, no. 1, (April 2006): 1–54.

Lumumba-Kasongo, Tukumbi. *Japan–Africa Relations*. New York: Palgrave Macmillan, 2010.

Mao, Yufeng. "China's Interests and Strategy in the Middle East and the Arab World." In *China and the Developing World: Beijing's Strategy for the Twenty-First Century*, edited by Joshua Eisenman, Eric Heginbotham, and Derek Mitchell, 118–119. Armonk, NY: Sharpe, 2007.

Marketos, Thrassy N. *China's Energy Geopolitics: The Shanghai Cooperation Organization and Central Asia*. New York: Routledge, 2009.

Masaki, Hisane. "Japan Energy: Goodbye Iran, Hello Iraq," *Asia Times Online*, November 7, 2006. www.atimes.com/atimes/Japan/HK07Dh01.html.

Matheswaran, M. "Air and Aerospace Partnership." In *China–Pakistan Strategic Cooperation: Indian Perspectives*, edited by Swaran Singh, 198. New Delhi, India: Manohar Publishers, 2007.

McAvoy, Audrey. "Adm.: Spratly Row Threatens Security, Stability," *Associated Press*, January 18, 2012.

McCaffrey, Barry. "Iran, Nukes, & Oil: The Gulf Confrontation." Powerpoint presentation presented at a seminar for NBC Executives and Producers, January 12, 2012.

Medeiros, Evan S. *China's International Behavior: Activism, Opportunism, and Diversification*. Santa Monica, CA: RAND Corporation, 2009.

Ministry of Defence, Government of India. *Annual Report 2011–2012*, New Delhi, India: Ministry of Defence, 2012.

Ministry of Defense, Japan. *Defense of Japan 2012*. Tokyo, Japan: Ministry of Defense, 2012.

Ministry of Foreign Affairs, Japan. *Diplomatic Bluebook 2012*. Tokyo, Japan: Ministry of Foreign Affairs, April 2012.

Mohan, C. Raja. *Samudra Manthan: Sino-Indian Rivalry in the Indo-Pacific*. Washington, DC: Carnegie Endowment for International Peace, 2012.

Moran, Daniel, and James A. Russell. "Introduction: the militarization of energy security." In *Energy Security and Global Politics: the Militarization of Resource Management*, edited by Daniel Moran and James A. Russell, 7. New York: Routledge, 2009.

Morrissey, Oliver, and Evious Zgovu. *The Impact of China and India on Sub-Saharan African Countries: Opportunities, Challenges and Policies*. London, UK: Commonwealth Secretariat, 2011.

Naidu, Sanusha. "India's African Relations: In the Shadow of China?" In *The Rise of China and India in Africa: Challenges, Opportunities and Critical Interventions*, edited by Fantu Cheru and Cyril Obi, 35, 39–41. New York: Zed Books, 2010.

Niazi, Tarique. "Gwadar: China's Naval Outpost on the Indian Ocean." *China Brief*, Vol. 5, Issue 4, February 14, 2005.

Nojonen, Matti. "Introduction: Adjusting to the Great Power Transition." In *Russia–China Relations: Current State, Alternative Futures, and Implications for the West*, edited by Arkady Moshes and Matti Nojonen, 14. Helsinki, Finland: Finnish Institute of Foreign Affairs, 2011.

Obi, Cyril. "African Oil in the Energy Security Calculations of China and India." In *The Rise of China and India in Africa: Challenges, Opportunities and Critical Interventions*, edited by Fantu Cheru and Cyril Obi, 181, 184, 185, 187. New York: Zed Books, 2010.

Oresman, Matthew. "Repaving the Silk Road: China's Emergence in Central Asia." In *China and the Developing World: Beijing's Strategy for the Twenty-First Century*, edited by Joshua Eisenman, Eric Heginbotham, and Derek Mitchell, 61–62, Armonk, NY: Sharpe, 2007.

Organization for Economic Co-operation and Development, *The Export Credits Arrangement*, 1978–2008. http://www.oecd.org/tad/xcred/40594872.pdf.

O'Rourke, Ronald. *China Naval Modernization: Implications for U.S. Naval Capabilities—Background and Issues for Congress*. Congressional Research Service Report RL33153, Washington, DC: Library of Congress, December 1, 2010.

Pehrson, Christopher J. *String of Pearls: Meeting the Challenge of China's Rising Power Across the Asian Littoral*. Carlisle, PA: Strategic Studies Institute, July 2006.

Pham, J. Peter. "China's African Strategy and Its Implications for U.S. Interests." *American Foreign Policy Interests*, Issue 28 (2006): 247.

Prabhakar, W. Lawrence. "The Maritime Convergence." In *China–Pakistan Strategic Cooperation: Indian Perspectives*, edited by Swaran Singh, 238. New Delhi, India: Manohar Publishers, 2007.

Rautava, Jouko. "Russia's Economic Policy and Russia–China Economic Relations." In *Russia–China Relations: Current State, Alternative Futures, and Implications for the West*, edited by Arkady Moshes and Matti Nojonen, 45–46. Helsinki, Finland: Finnish Institute of Foreign Affairs, 2011.

Roett, Riordan, and Guadalupe Paz. "Introduction: Assessing the Implications of China's Growing Presence in the Western Hemisphere." In *China's Expansion Into the Western Hemisphere: Implications for Latin America and the United States*, edited by Riordan Roett and Guadalupe Paz, 16. Washington, DC: Brookings Institution Press, 2008.

Root, Hilton. *Alliance Curse: How America Lost the Third World.* Washington, DC: Brookings Institution Press, May 1, 2008.

Rotberg, Robert I. "China's Quest for Resources, Opportunities, and Influence in Africa." In *China Into Africa: Trade, Aid and Influence,* edited by Robert I. Rotberg, 1. Washington, DC: Brookings Institution Press, 2008.

Sakhuja, Vijay. "Naval Cooperation." In *China–Pakistan Strategic Cooperation: Indian Perspectives,* edited by Swaran Singh, 215–220. New Delhi, India: Manohar Publishers, 2007.

Shambaugh, David. "China eyes Europe in the world: real convergence or cognitive dissonance?" In *China–Europe Relations: Perceptions, Policies and Prospects,* edited by David Shambaugh, Eberhard Sandschneider, and Zhou Hong, 136. New York: Routledge, 2008.

Shinn, David H. "Military and Security Relations: China, Africa, and the Rest of the World." In *China Into Africa: Trade, Aid and Influence,* edited by Robert I. Rotberg, 154, 162, 165, 177, 180, 183. Washington, DC: Brookings Institution Press, 2008.

Small, Andrew. "The US Factor." In *China–EU: A Common Future,* edited by Stanley Crossick and Etienne Reuter, 74. Singapore: World Scientific Publishing, 2007.

Stanzel, Volker. "The EU and China in the Global System." In *China–Europe Relations: Perceptions, Policies and Prospects,* edited by David Shambaugh, Eberhard Sandschneider and Zhou Hong, 258. New York: Routledge, 2008.

Sutter, Robert G. *Chinese Foreign Relations: Power and Policy Since the Cold War.* Lanham, MD: Rowman & Littlefield, 2008.

Sutter, Robert G. "China's Rise, Southeast Asia, and the United States: Is A China-Centered Order Marginalizing the United States." In *China, United States, and Southeast Asia: Contending Perspectives on Politics, Security, and Economics,* edited by Evelyn Goh and Sheldon W. Simon, 92. New York: Routledge, 2008.

Taylor, Ian. *China's New Role in Africa.* London, UK: Rienner, 2009.

Tellis, Ashley J. "China and India in Asia." In *The India–China Relationship: What the United States Needs to Know,* edited by Francine R. Frankel and Harry Harding, 134–177. New York: Columbia University Press, 2004.

Teng, Chung-Chian. "Hegemony or Partnership: China's Strategy and Diplomacy Toward Latin America." In *China and the Developing World: Beijing's Strategy for the Twenty-First Century,* edited by Joshua Eisenman, Eric Heginbotham, and Derek Mitchell, 97. Armonk, NY: Sharpe, 2007.

U.S. Department of Defense, *Annual Report to Congress: Military and Security Developments Involving the People's Republic of China,* 2011.

———. *Proliferation: Threat and Response,* 1996.

U.S. House, Congressional Budget Office. *China's Growing Demand for Oil and Its Impact on U.S. Petroleum Markets,* Pub. 2561, April 2006, 1.

U.S. House. Subcommittee on the Western Hemisphere, *China's Influence on the Western Hemisphere,* Hearing, April 6, 2005 (Serial No. 109–63). Washington: Government Printing Office, 2005.

U.S. Library of Congress, Congressional Research Service, *China's Foreign Policy and "Soft Power" in South America, Asia, and Africa: A Study Prepared for the Committee on Foreign Relations, United States Senate.* Washington, DC: U.S. Government Printing Office, 2008.

U.S. Senate, Committee on Foreign Relations, *The Emergence of China Throughout Asia: Security and Economic Consequences for the United States,* Hearing, June 7, 2005 (S. Hrg. 109–239). Washington: Government Printing Office, 2005.

Verachia, Abdullah. "Unpacking India's African Engagement," Powerpoint Presentation. Paris, France: Frontier Advisory (October 2010); 24–26.

Vines, Alex, and Indira Campos. "China and India in Angola." In *The Rise of China and India in Africa: Challenges, Opportunities and Critical Interventions,* edited by Fantu Cheru and Cyril Obi, 195, 200, 205. New York: Zed Books, 2010.

Watson, Cynthia, "The PLA in Latin America," *China Brief,* Volume 7, Issue 20, October 31, 2007.

Wenping, He. "The Darfur Issue: A New Test for China's Africa Policy." In *The Rise of China and India in Africa: Challenges, Opportunities and Critical Interventions,* edited by Fantu Cheru and Cyril Obi, 159. New York: Zed Books, 2010.

Xiang, Lanxin. "An Alternative Chinese View." In *China's Expansion Into the Western Hemisphere: Implications for Latin America and the United States,* edited by Riordan Roett and Guadalupe Paz, 55. Washington, DC: Brookings Institution Press, 2008.

Index

About the Author

George G. Eberling completed his Ph.D. in Political Science from Claremont Graduate University in May 2010. *Future Oil Demands of China, India, and Japan: Policy Scenarios and Implications* is his second publication. He previously published *Chinese Energy Futures and Their Implications for the United States* in October 2011.

Lightning Source UK Ltd.
Milton Keynes UK
UKOW05n1807220714

235564UK00001B/34/P

9 780739 191811